Full-Stack Web Development with Go

I0010601

Build your web applications quickly using the Go programming language and Vue.js

Nanik Tolaram

Nick Glynn

BIRMINGHAM—MUMBAI

Full-Stack Web Development with Go

Group Product Manager: Pavan Ramchandani

Publishing Product Manager: Bhavya Rao

Senior Content Development Editor: Feza Shaikh

Technical Editor: Saurabh Kadave

Copy Editor: Safis Editing

Project Coordinator: Manthan Patel

Proofreader: Safis Editing

Indexer: Tejal Daruwale Soni

Production Designer: Shyam Sundar Korumilli

Marketing Coordinator: Anamika Singh

First published: February 2023

Production reference: 1270123

Published by Packt Publishing Ltd.
Livery Place
35 Livery Street
Birmingham
B3 2PB, UK.

ISBN 978-1-80323-419-9

www.packtpub.com

To my dearest Mum, who always supported me in pursuing my dreams and encouraged me to keep on going no matter what life brought.

To my late Dad, who stood by me and encouraged me to write my very first book when I was 17 years old.

To my beautiful wife and best friend, for allowing me the time to write the book and supporting me in every step of our life.

To both my sons, Rahul and Manav, for allowing me to spend time in front of the computer on weekends to chase my dream and passion. Last but not least, to God, for giving me this life and the opportunity to be where I am in this world.

– Nanik Tolaram

I would like to thank my family for their love; my beautiful daughter Inara, for always being there to brighten my day; and my beautiful partner Kate, for all her support in my business efforts and in writing this book.

– Nick Glynn

Contributors

About the authors

Nanik Tolaram is a big proponent of open source software. He has dabbled in different programming languages, such as Java, JavaScript, C, and C++. He has developed different products from the ground up while working in start-up companies. He is a software engineer at heart, but he loves to write technical articles and share his knowledge with others. He learned to program with Go during the COVID-19 pandemic and hasn't looked back.

I want to thank the Packt team – Feza Shaikh, Bhavya Rao, Manthan Patel, and Mark D'Souza – for their guidance and patience in helping us complete the book.

Nick Glynn is the founder and current chief product officer of FeeWise, a US payments and finance platform.

He has previously worked in CTO, senior, and principal engineering roles developing products, training, and consulting for companies such as Intel, Qualcomm, Hewlett Packard, L3, and many more.

With a broad range of experience from board bring-up, Linux driver development, and systems development up through to building and deploying platforms that power investment and financial institutions, Nick is always looking to build pragmatic solutions to real-world problems.

Nick also continues his independent efforts as a trainer and consultant, delivering courses and expertise globally through his company Curiola (`www.curiola.com`).

About the reviewers

Pablo David Garaguso was born in Mar del Plata, Argentina. He graduated with two degrees in computer sciences and enterprise systems from CAECE University and later on received an MBA from CEMA University in Buenos Aires, Argentina. He has worked as an attached professor, project leader, and international manager, and currently occupies a position as a solutions architect in R&D for FamilySearch Int. Europe. Based in Finland, he has published a variety of books according to his many interests, from novels and fiction to software engineering. His latest book, *Vue 3 Applications and Design Patterns*, will be published by Packt in 2023.

Suraj Bobade is an experienced software professional, currently located in Pune, India. He completed a B.Tech in computer science from Walchand College of Engineering, Sangli.

He is passionate about software development with a keen interest in product management. He builds user-first feature-rich products while driving critical software and product design decisions.

Go has been his go-to choice for building the microservice backend and prototypes. Considering the simplicity and increasing adoption by the open source community, Suraj is optimistic that readers of this book will learn in-demand skills.

Tan Quach is an experienced software engineer with a career spanning over 25 years and exotic locations such as London, Canada, Bermuda, and Spain. He has worked with a wide variety of languages and technologies for companies such as Deutsche Bank, Merrill Lynch, and Progress Software, and loves diving deep into experimenting with new ones.

Tan's first foray into Go began in 2017 with a proof-of-concept application built over a weekend and productionized and released 3 weeks later. Since then, Go has been his language of choice when starting any project.

When he can be torn away from the keyboard, Tan enjoys cooking meat over hot coals and open flames and making his own charcuterie.

Nima Yahyazadeh is a Software Architect focused on developing solutions for startups. He has years of experience developing distributed and cloud-native solutions. He has worked at medium to large corporations such as Amazon Web Services, Intel, and Appian. He is currently the founder and CEO of a consulting company, Lenzo LLC, that has helped more than five startups to architect, develop, and deliver exciting features to their customers. He is passionate about AWS, Kubernetes, Elasticsearch, Kafka, and Golang.

Table of Contents

Part 2: Serving Web Content

4

Serving and Embedding HTML Content   63

5

Securing the Backend and Middleware   85

6

Moving to API-First   99

Part 3: Single-Page Apps with Vue and Go

7

Frontend Frameworks   119

8

Frontend Libraries   135

9

Tailwind, Middleware, and CORS   151

10

Session Management   169

Part 4: Release and Deployment

11

Feature Flags   185

12

Building Continuous Integration   199

13

Dockerizing an Application 219

14

Cloud Deployment 237

Index 271

Other Books You May Enjoy 274

Preface

Full-Stack Web Development with Go walks you through creating and developing a complete modern web service, from authn/authz, interop, server-side rendering, and databases, to modern frontend frameworks with Tailwind and Go-powered APIs, complete with step-by-step explanations of essential concepts, practical examples, and self-assessment questions. The book will begin by looking at how to structure the app and look at the relevant pieces, such as database and security, before integrating all the different parts together to build a complete web product.

Who this book is for

Developers with experience of a mixture of frontend and backend development will be able to put their knowledge to work with the practical guide in this book. The book will give them the know-how to glue together their skills and allow them to build a complete stack web application.

What this book covers

Chapter 1, Building the Database and Model, looks at building our database for our sample application. We will also explore different ways to communicate with databases using Golang.

Chapter 2, Application Logging, considers how designing an application requires examining it internally without going through reams of code, and the only way to do this is by logging. We will learn how to do this by running a centralized logger that will host all of our logging information. We will also learn how to log from inside our application.

Chapter 3, Application Metrics and Tracing, considers how having logging applied inside our application will assist in troubleshooting issues when the application is running. The other thing that helps is information about the interaction of the different components inside our application, which we will also look at in this chapter.

Chapter 4, Serving and Embedding HTML Content, sees us begin work on implementing the REST endpoints needed for our financial application. The first version of the app will show simple content rendered by the backend.

Chapter 5, Securing the Backend and Middleware, notes that we need to secure our application so that we can ensure users see only the data that they should. We will discuss some of the ways we can protect our endpoints using cookies, session management, and other types of middleware available.

Chapter 6, Moving to API - First, starts by laying the groundwork for frontend applications to consume our data. We'll introduce marshaling/unmarshaling data into custom structures and see how to set up JSON-consuming endpoints and use cURL to verify.

Chapter 7, Frontend Frameworks, discusses the state of web development, introduces the React and Vue frameworks, and sees us employ them to create a simple app that's similar to our previous one.

Chapter 8, Frontend Libraries, examines how to leverage tools and libraries to help us, as full stack developers, work fast!

Chapter 9, Tailwind, Middleware, and CORS, has us securing our app and getting it talking to our Go-powered backend.

Chapter 10, Session Management, focuses on session management while introducing state management with Vuex and how to structure apps based on user permissions.

Chapter 11, Feature Flags, introduces feature flags (sometimes called *feature toggles*) as a technique deployed to enable/disable certain features of the application as required, depending on a given condition. For example, if a newly deployed application containing a new feature has a bug and we know it will take time for the bug to be fixed, the decision can be made to turn off the feature without deploying any new code to do so.

Chapter 12, Building Continuous Integration, notes that while building applications is a big part of the puzzle, we need to be sure that what the team builds can be validated and tested properly. This is where continuous integration comes in. Having a proper continuous integration process is super important to ensure that everything deployed to production has been tested, verified, and checked securely.

Chapter 13, Dockerizing an Application, notes that while developing an application is one side of the coin, the other side is to make sure that it can be deployed and used by our end user. To make deployment simpler, we can package applications such that they can be run inside a container. Operationally, this allows applications to be deployed in the cloud from anywhere.

Chapter 14, Cloud Deployment, shows how deploying applications to a cloud environment is the last step in delivering features for the end user to use. Cloud deployment is complex and sometimes quite specific to particular cloud vendors. In this chapter, we will focus on deploying applications into the AWS cloud infrastructure.

To get the most out of this book

You will need the following installed on your computer: Node.js (version 16 or above), Docker (or Docker Desktop), the Golang compiler, and an IDE such as GoLand or VSCode.

Software/hardware covered in the book	Operating system requirements
Golang 1.16 and above	macOS, Linux, Windows (via WSL2)
Docker	macOS, Linux, Windows (via WSL2)
An IDE (VSCode or GoLand)	macOS, Linux, Windows

If you are using the digital version of this book, we advise you to type the code yourself or access the code from the book's GitHub repository (a link is available in the next section). Doing so will help you avoid any potential errors related to the copying and pasting of code.

Download the example code files

You can download the example code files for this book from GitHub at `https://github.com/PacktPublishing/Full-Stack-Web-Development-with-Go`. If there's an update to the code, it will be updated in the GitHub repository.

Download the color images

We also provide a PDF file that has color images of the screenshots and diagrams used in this book. You can download it here: `https://packt.link/EO4sG`.

Conventions used

There are a number of text conventions used throughout this book.

`Code in text`: Indicates code words in text, database table names, folder names, filenames, file extensions, pathnames, dummy URLs, user input, and Twitter handles. Here is an example: "We call `next.ServerHTTP(http.ResponseWriter, *http.Request)` to continue and indicate successful handling of a request."

A block of code is set as follows:

```
go func() {
  ...
  s.SetAttributes(attribute.String("sleep", "done"))
  s.SetAttributes(attribute.String("go func", "1"))
  ...
}()
```

Any command-line input or output is written as follows:

```
[INFO] 2021/11/26 21:11 This is an info message, with colors
(if the output is terminal)
```

Bold: Indicates a new term, an important word, or words that you see onscreen. For instance, words in menus or dialog boxes appear in **bold**. Here is an example: "You will get a **Login unsuccessful** message."

> **Tips or important notes**
> Appear like this.

Get in touch

Feedback from our readers is always welcome.

General feedback: If you have questions about any aspect of this book, email us at `customercare@packtpub.com` and mention the book title in the subject of your message.

Errata: Although we have taken every care to ensure the accuracy of our content, mistakes do happen. If you have found a mistake in this book, we would be grateful if you would report this to us. Please visit `www.packtpub.com/support/errata` and fill in the form.

Piracy: If you come across any illegal copies of our works in any form on the internet, we would be grateful if you would provide us with the location address or website name. Please contact us at `copyright@packt.com` with a link to the material.

If you are interested in becoming an author: If there is a topic that you have expertise in and you are interested in either writing or contributing to a book, please visit `authors.packtpub.com`.

Share Your Thoughts

Once you've read, we'd love to hear your thoughts! Scan the QR code below to go straight to the Amazon review page for this book and share your feedback.

https://packt.link/r/1803234199

Your review is important to us and the tech community and will help us make sure we're delivering excellent quality content.

Download a free PDF copy of this book

Thanks for purchasing this book!

Do you like to read on the go but are unable to carry your print books everywhere? Is your eBook purchase not compatible with the device of your choice?

Don't worry, now with every Packt book you get a DRM-free PDF version of that book at no cost.

Read anywhere, any place, on any device. Search, copy, and paste code from your favorite technical books directly into your application.

The perks don't stop there, you can get exclusive access to discounts, newsletters, and great free content in your inbox daily

Follow these simple steps to get the benefits:

1. Scan the QR code or visit the link below

https://packt.link/free-ebook/9781803234199

2. Submit your proof of purchase
3. That's it! We'll send your free PDF and other benefits to your email directly

Part 1:
Building a Golang Backend

In *Part 1*, we will build the backend components for our sample application. We will build the database with the models for the Go backend of our application. We will also build secure REST API endpoints that will have logging and tracing functionalities.

This part includes the following chapters:

1
Building the Database and Model

In this chapter, we will design the database that our sample application will use. We will walk through the design of the database and look at some of the tools that we are going to use to help us on our database design journey. We will be using the **Postgres** database and will look at how to run it locally using **Docker**. What is Docker? In simple terms, Docker is a tool that allows developers to run a variety of applications such as the database, the HTTP server, system tools, and so on – locally or in the cloud. Docker removes the need to install all the different dependencies required to use a particular application such as a database, and it makes it easier to manage and maintain applications than installing on bare metal in both local and cloud environments. This is possible using Docker because it packages everything into a single file similar to how a compressed file contains different files internally.

We will learn how to design a database that supports the features that we want to build, such as the following:

- Creating an exercise
- Creating a workout plan
- Logging in to the system

We will also explore tools that will help in automatic code generation based on SQL queries, which reduces the amount of database-related code that needs to be written to a large extent. Readers will learn to use the tool to also auto-generate all the relevant CRUD operations without writing a single line of Go code.

In this chapter, we'll be covering the following:

- Installing Docker
- Setting up Postgres
- Designing the database

- Installing sqlc
- Using sqlc
- Setting up the database
- Generating CRUD with sqlc
- Building the makefile

Technical requirements

In this book, we will be using version 1.16 of the Go programming language, but you are free to use later versions of Go, as the code will work without any changes. To make it easy, all the relevant files explained in this chapter can be checked out at `https://github.com/PacktPublishing/Full-Stack-Web-Development-with-Go/tree/main/Chapter01`. To work on the sample code in this chapter, make sure you change the directory to `Chapter 1 – Full-Stack-Web-Development-with-Go/chapter1`. If you are using Windows as a development machine, use WSL2 to perform all the different operations explained in this chapter.

Installing Docker

In this book, we will be using Docker to do things such as running databases and executing database tools, among others. You can install either Docker Desktop or Docker Engine. To understand more about the difference between Docker Desktop and Engine, visit the following link: `https://docs.docker.com/desktop/linux/install/#differences-between-docker-desktop-for-linux-and-docker-engine`. The authors use Docker Engine in Linux and Docker Desktop for Mac.

If you are installing Docker Desktop on your local machine, the following are the links for the different operating systems:

- Windows – `https://docs.docker.com/desktop/windows/install/`
- Linux – `https://docs.docker.com/desktop/linux/install/`
- macOS – `https://docs.docker.com/desktop/mac/install/`

If you want to install Docker binaries, you can follow the following guide: `https://docs.docker.com/engine/install/binaries/`.

Setting up Postgres

The database we chose for the sample application is Postgres; we chose Postgres over other databases because of the wide variety of open source tools available for building, configuring, and maintaining Postgres. Postgres has been open source from version 1 since 1989 and it is used by big tech startups worldwide. The project has a lot of community support in terms of tools and utilities, which makes it easier to manage and maintain. The database is suitable for small all the way to big replicated data stores.

The easiest way to run it locally is to run it as a Docker container. First, use the following command to run Postgres:

```
docker run --name test-postgres \
 -e POSTGRES_PASSWORD=mysecretpassword -p 5432:5432 -d postgres
```

The command will run `postgres` on port `5432`; if by any chance you have other applications or other Postgres instances listening to this port, the command will fail. If you need to run Postgres on a different port, change the `-p` parameter (for example, `-p 5555:5432`) to a different port number.

If successful, you will see the container ID printed out. The ID will differ from what is shown here:

```
f7bdfb7d2c10c5f0c9227c9b0a720f21d3c7fa65907eb-
0c546b8f20f12621102
```

Check whether Postgres is up and running by using `docker ps`. The next thing to do is use the `psql-client` tool to connect to Postgres to test it out. A list of the different Postgres client tools available on different platforms can be found here: `https://wiki.postgresql.org/wiki/PostgreSQL_Clients`.

We will use the standard `postgres psql` tool using Docker. Open another terminal and use the following Docker command to run `psql`:

```
docker exec -it test-postgres psql -h localhost -p 5432 -U
postgres -d postgres
```

What we are doing is executing the `psql` command inside the running Postgres container. You will see output such as the following, indicating that it has successfully connected to the Postgres database:

```
psql (12.3, server 14.5 (Debian 14.5-1.pgdg110+1))
WARNING: psql major version 12, server major version 14.
         Some psql features might not work.
Type "help" for help.


postgres=#
```

On a successful connection, you will see the following output. Note that the warning message mentions server major version 14 – this is to indicate that the server version is newer than the current `psql` version as per the documentation (`https://www.postgresql.org/docs/12/app-psql.html`). The `psql` client will work without any problem with the Postgres server:

```
psql (12.3, server 14.0 (Debian 14.0-1.pgdg110+1))
WARNING: psql major version 12, server major version 14.
         Some psql features might not work.
Type "help" for help.

postgres=#
```

Exit `psql` to go back to the command prompt by typing `exit`.

The following is some guidance on common errors when trying to connect to the database:

Error Message	Description
`psql: error: could not connect to server: FATAL:  password authentication failed for user "postgres"`	The password specified when running Postgres does not match with the password passed in using `psql`. Check the password.
psql: error: could not connect to server: could not connect to server: Host is unreachable	The IP address that you use to connect to Postgres is wrong.

With this, you have completed the local setup of Postgres and are now ready to start looking into designing the database.

Designing the database

In this section, we will look at how to design the database to allow us to store information for the fitness tracking application. The following screenshot shows a mockup of the application:

Figure 1.1 – Screenshot of the sample application

Looking at these functionalities, we will look at designing a database structure that will look like the following entity relationship diagram:

> **Entity relationship diagram**
> An entity relationship diagram shows the relationships of entity sets stored in a database.

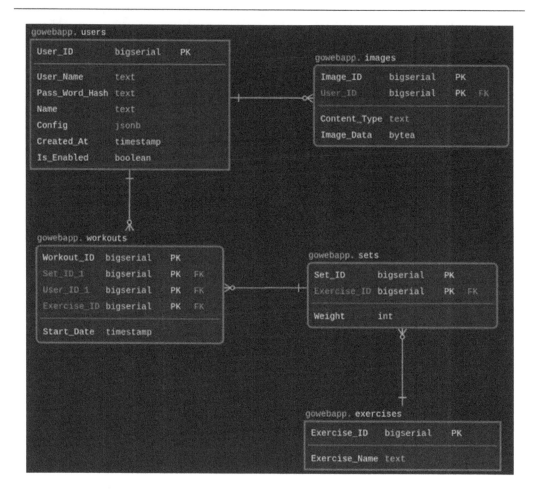

Figure 1.2 – Entity relationship diagram of our fitness application

Let's drill further into each table to understand the data that they contain:

Table Name	Description
Users	This table contains user information for login purposes. Passwords will be stored as a hash, not plaintext.
Images	This table contains images of exercises that users want to do. This table will store all the exercise images that the user uploads.
Exercises	This table contains the name of the exercise that the user wants to do. Users will define what kind of exercise they want to do.
Sets	This table contains the number of sets of each exercise that the user wants to do.
Workouts	This table contains the workouts that the user wants to do. Users define a workout as a combination of exercises with the number of sets that they want to do.

The trade-off we are making to store images in the database is to simplify the design; in reality, this might not be suitable for bigger images and production. Now that we have defined the database structure and understand what kind of data it will store, we need to look at how to implement it. One of the major criteria that we want to focus on is to completely separate writing SQL from the code; this way, we have a clear separation between the two, which will allow higher maintainability.

Installing sqlc

We have defined the database structure so now it's time to talk a bit more about the tool that we are going to be using called sqlc. **sqlc** is an open source tool that generates type-safe code from SQL; this allows developers to focus on writing SQL and leave the Go code to sqlc. This reduces the development time, as sqlc takes care of the mundane coding of queries and types.

The tool is available at `https://github.com/kyleconroy/sqlc`. The tool helps developers focus on writing the SQL code that is needed for the application and it will generate all the relevant code needed for the application. This way, developers will be using pure Go code for database operations. The separation is clean and easily trackable.

The following diagram shows the flow that developers normally adopt when using the tool at a high level.

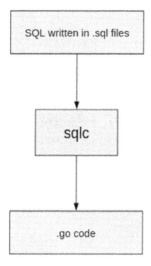

Figure 1.3 – Flow to use sqlc to generate Go code

All SQL code will be written in `.sql` files, which will be read and converted by the sqlc tool into the different Go code.

Download and install SQL binary by using the following command:

```
go install github.com/kyleconroy/sqlc/cmd/sqlc@latest
```

Make sure your path includes the GOPATH/bin directory – for example, in our case, our path looks like the following:

```
…:/snap/bin:/home/nanik/goroot/go1.16.15/go/bin:/home/nanik/go/
bin
```

If you don't have GOPATH as part of the PATH environment variable, then you can use the following command to run sqlc:

```
$GOPATH/bin/sqlc
Usage:
  sqlc [command]


Available Commands:
  compile      Statically check SQL for syntax and type
  errors
  completion   Generate the autocompletion script for the
  specified shell
  generate     Generate Go code from SQL
  help         Help about any command
  init         Create an empty sqlc.yaml settings file
  upload       Upload the schema, queries, and configuration
  for this project
  version      Print the sqlc version number


Flags:
  -x, --experimental    enable experimental features (default:
  false)
  -f, --file string     specify an alternate config file
  (default: sqlc.yaml)
  -h, --help            help for sqlc
```

Use "sqlc [command] --help" for more information about a command.

At the time of writing, the latest version of sqlc is v1.13.0.

Now that we have installed the tool and understand the development workflow that we will be following when using the tool, we will look at how to use the tool for our application.

Using sqlc

First, let's take a look at the different commands provided by sqlc and how they work.

Commands	Explanation
compile	This command helps check SQL syntax and reports any typing errors.
completion	This command is to generate an auto-completion script for your environment. The following are the supported environments: Bash, Fish, PowerShell, and zsh.
generate	A command to generate the .go files based on the provided SQL statements. This will be the command that we will be using a lot for the application.
init	This command is the first command that is used to initialize your application to start using this tool.

The following will show how to get started with using sqlc to set up a project. Create a directory inside chapter1 – for example, dbtest – and change the directory to the new directory (dbtest). Next, we will run sqlc with the init command:

```
sqlc init
```

This will automatically generate a file called sqlc.yaml, which contains a blank configuration as shown here:

```
version: "1"
project:
    id: ""
packages: []
```

The sqlc.yaml contains configuration information that sqlc will use to generate all the relevant .go code for our SQL statements.

Let's take a look at the structure of the .yaml file to understand the different properties. The following shows an example of a completed structure:

```
version: "1"
packages:
  - name: "db"
    path: "db"
    queries: "./sqlquery"
    schema: "./sqlquery/schema/"
```

```
engine: "postgresql"
sql_engine: "database/sql"
emit_db_tags: "true"
emit_prepared_queries: true
emit_interface: false
emit_exact_table_names: false
emit_empty_slices: false
emit_exported_queries: false
emit_json_tags: true
json_tags_case_style: "snake"
output_db_file_name: "db.go"
output_models_file_name: "dbmodels.go"
output_querier_file_name: "dbquerier.go"
output_files_suffix: "_gen"
```

The following table explains the different fields:

Tag Name	Description
Name	Any string to be used as the package name.
Path	Specifies the name of the directory that will host the generated .go code.
Queries	Specifies the directory name containing the SQL queries that sqlc will use to generate the .go code.
Schema	A directory containing SQL files that will be used to generate all the relevant .go files.
Engine	Specifies the database engine that will be used: sqlc supports either MySQL or Postgres.
emit_db_tags	Setting this to true will generate the struct with db tags – for example: type ExerciseTable struct { ExerciseID int64 `db:"exercise_id"     ExerciseName string `db:"exercise_name" }
emit_prepared_queries	Setting this to true instructs sqlc to support prepared queries in the generated code.

`emit_interface`	Setting this to `true` will instruct sqlc to generate the querier interface.
`emit_exact_table_names`	Setting this to `true` will instruct sqlc to mirror the struct name to the table name.
`emit_empty_slices`	Setting this to `true` will instruct sqlc to return an empty slice for returning data on many sides of the table.
`emit_exported_queries`	Setting this to `true` will instruct sqlc to allow the SQL statement used in the auto-generated code to be accessed by an outside package.
`emit_json_tags`	Setting this to `true` will generate the struct with JSON tags.
`json_tags_case_style`	This setting can accept the following – `camel`, `pascal`, `snake`, and `none`. The case style is used for the JSON tags used in the struct. Normally, this is used with `emit_json_tags`.
`output_db_file_name`	Name used as the filename for the auto-generated database file.
`output_models_file_name`	Name used as the filename for the auto-generated model file.
`output_querier_file_name`	Name used as the filename for the auto-generated querier file.
`output_files_suffix`	Suffix to be used as part of the auto-generated query file.

We have looked at the different parameters available in the tool, along with how to use the `.yaml` file to specify the different properties used to generate the relevant Go files. In the next section, we will set up our sample app database.

Setting up the database

We need to prepare and create the database using the `psql` client tool. The SQL database script can be found inside `schema.sql` under the `db` folder in the GitHub repository, and we are going to use this to create all the relevant tables inside Postgres.

Change the directory to `chapter1` and run the Postgres database using the following Docker command:

```
docker run --name test-postgres -e POSTGRES_
PASSWORD=mysecretpassword -v $(pwd):/usr/share/chapter1 -p
5432:5432 postgres
```

Once `postgres` is running, use the following command to enter into `psql`:

```
docker exec -it test-postgres psql -h localhost -p 5432 -U
postgres -d postgres
```

Once inside the `psql` command, run the following:

```
\i /usr/share/chapter1/db/schema.sql
```

This will instruct `psql` to execute the commands inside `schema.sql`, and on completion, you will see the following output:

```
postgres=# \i /usr/share/chapter1/db/schema.sql
CREATE SCHEMA
CREATE TABLE
CREATE TABLE
CREATE TABLE
CREATE TABLE
CREATE TABLE
```

To reconfirm that everything is set up correctly, use the following command (do not forget to include the dot after `gowebapp`):

```
\dt gowebapp.*
```

You should see the following output:

```
postgres=# \dt gowebapp.*
            List of relations
   Schema   |    Name    | Type  |  Owner
------------+------------+-------+----------
 gowebapp   | exercises  | table | postgres
 gowebapp   | images     | table | postgres
 gowebapp   | sets       | table | postgres
 gowebapp   | users      | table | postgres
 gowebapp   | workouts   | table | postgres
(5 rows)
```

Now that we have completed setting up our database, we are ready to move to the next section, where we will be setting up sqlc to generate the Go files.

Generating CRUD with sqlc

CRUD stands for **Create, Read, Update, and Delete**, which refers to all the major functions that are inherent to relational databases. In this section, we will do the following for the application:

- Complete the sqlc configuration file
- Create SQL query files

Once done, we will be able to autogenerate the different files required to allow us to perform CRUD operations to the database from the application. First, open `sqlc.yaml` and enter the following configuration:

```
---
version: '1'
packages:
  - name: chapter1
    path: gen
    schema: db/
    queries: queries/
    engine: postgresql
    emit_db_tags: true
    emit_interface: false
    emit_exact_table_names: false
    emit_empty_slices: false
    emit_exported_queries: false
    emit_json_tags: true
    json_tags_case_style: camel
    output_files_suffix: _gen
    emit_prepared_queries: false
```

Our application is now complete with all that we need for the database, and sqlc will autogenerate the `.go` files. The following is how the application directory and files will look:

```
.
├── db
│   └── schema.sql
├── go.mod
├── queries
│   └── query.sql
└── sqlc.yaml
```

We can run sqlc to generate the `.go` files using the following command:

```
sqlc generate
```

By default, sqlc will look for the `sqlc.yaml` file. If the filename is different, you can specify it using the `-f` flag as follows:

```
sqlc generate -f sqlc.yaml
```

Once the operation completes, there will be no output; however, a new directory called `gen` will be generated as shown here:

```
./gen/
├── db.go
├── models.go
└── query.sql_gen.go
```

We have completed the auto-generation process using sqlc; now, let's take a look at the schema and queries that sqlc uses to generate the code.

The following is a snippet of the `schema.sql` file that is used by sqlc to understand the structure of the database:

```
CREATE SCHEMA IF NOT EXISTS gowebapp;

CREATE TABLE gowebapp.users (
User_ID         BIGSERIAL PRIMARY KEY,
User_Name       text NOT NULL,
....
);
....

CREATE TABLE gowebapp.sets (
Set_ID       BIGSERIAL PRIMARY KEY,
Exercise_ID BIGINT NOT NULL,
Weight       INT NOT NULL DEFAULT 0
);
```

The other file sqlc uses is the query file. The query file contains all the relevant queries that will perform CRUD operations based on the database structure given here. The following is a snippet of the `query.sql` file:

```
-- name: ListUsers :many
-- get all users ordered by the username
SELECT *
```

```
FROM gowebapp.users
ORDER BY user_name;
...
-- name: DeleteUserImage :exec
-- delete a particular user's image
DELETE
FROM gowebapp.images i
WHERE i.user_id = $1;
...
-- name: UpsertExercise :one
-- insert or update exercise of a particular id
INSERT INTO gowebapp.exercises (Exercise_Name)
VALUES ($1) ON CONFLICT (Exercise_ID) DO
UPDATE
    SET Exercise_Name = EXCLUDED.Exercise_Name
    RETURNING Exercise_ID;

-- name: CreateUserImage :one
-- insert a new image
INSERT INTO gowebapp.images (User_ID, Content_Type,
                             Image_Data)
values ($1,
        $2,
        $3) RETURNING *;
...
```

Using query.sql and schema.sql, sqlc will automatically generate all the relevant .go files, combining information for these two files together and allowing the application to perform CRUD operations to the database by accessing it like a normal struct object in Go.

The last piece that we want to take a look at is the generated Go files. As shown previously, there are three auto-generated files inside the gen folders: db.go, models.go, and query.sql_gen.go.

Let's take a look at each one of them to understand what they contain and how they will be used in our application:

- db.go:

  This file contains an interface that will be used by the other auto-generated files to make SQL calls to the database. It also contains functions to create a Go struct that is used to do CRUD operations.

A new function is used to create a query struct, passing in a DBTX struct. The DBTX struct implementation is either `sql.DB` or `sql.Conn`.

The `WithTx` function is used to wrap the `Queries` object in a database transaction; this is useful in situations where there could be an update operation on multiple tables that need to be committed in a single database transaction:

```go
func New(db DBTX) *Queries {
  return &Queries{db: db}
}


func (q *Queries) WithTx(tx *sql.Tx) *Queries {
  return &Queries{
    db: tx,
  }
}
```

- `models.go`:

  This file contains the struct of the tables in the database:

```go
type GowebappExercise struct {
  ExerciseID    int64   `db:"exercise_id"
    json:"exerciseID"`
  ExerciseName string `db:"exercise_name"
    json:"exerciseName"`
}
...

type GowebappWorkout struct {
  WorkoutID int64      `db:"workout_id"
    json:"workoutID"`
  UserID     int64      `db:"user_id" json:"userID"`
  SetID      int64      `db:"set_id" json:"setID"`
  StartDate time.Time `db:"start_date"
    json:"startDate"`
}
```

- `query.sql_gen.go`:

This file contains CRUD functions for the database, along with the different parameters struct that can be used to perform the operation:

```go
const deleteUsers = `-- name: DeleteUsers :exec
DELETE FROM gowebapp.users
WHERE user_id = $1
`

func (q *Queries) DeleteUsers(ctx context.Context,
userID int64) error {
  _, err := q.db.ExecContext(ctx, deleteUsers, userID)
  return err
}

...

const getUsers = `-- name: GetUsers :one
SELECT user_id, user_name, pass_word_hash, name, config,
created_at, is_enabled FROM gowebapp.users
WHERE user_id = $1 LIMIT 1
`

func (q *Queries) GetUsers(ctx context.Context, userID
int64) (GowebappUser, error) {
  row := q.db.QueryRowContext(ctx, getUsers, userID)
  var i GowebappUser
  err := row.Scan(
          &i.UserID,
          &i.UserName,
          &i.PassWordHash,
          &i.Name,
          &i.Config,
          &i.CreatedAt,
          &i.IsEnabled,
  )
  return i, err
```

```
  }
  ...
```

Now that the database and auto-generated data to perform CRUD operations are complete, let's try all this by doing a simple insert operation into the user table.

The following is a snippet of main.go:

```go
package main

import (
  ...
)

func main() {
  ...

  // Open the database
  db, err := sql.Open("postgres", dbURI)
  if err != nil {
    panic(err)
  }

  // Connectivity check
  if err := db.Ping(); err != nil {
    log.Fatalln("Error from database ping:", err)
  }

  // Create the store
  st := chapter1.New(db)

  st.CreateUsers(context.Background(),
  chapter1.CreateUsersParams{
    UserName:     "testuser",
    PassWordHash: "hash",
    Name:         "test",
  })
}
```

The app is doing the following:

1. Initializing the URL and opening the database

2. Pinging the database

3. Creating a new user using the `CreateUsers(..)` function

Make sure you are in the `chapter1` directory and build the application by running the following command:

```
go build -o chapter1
```

The compiler will generate a new executable called `chapter1`. Execute the file, and on a successful run, you will see the data inserted successfully into the `users` table:

```
2022/05/15 16:10:49 Done!
Name : test, ID : 1
```

We have completed setting up everything from the database and using sqlc to generate the relevant Go code. In the next section, we are going to put everything together for ease of development.

Building the makefile

A makefile is a file that is used by the `make` utility; it contains a set of tasks consisting of different combined shell scripts. Makefiles are most used to perform operations such as compiling source code, installing executables, performing checks, and many more. The `make` utility is available for both macOS and Linux, while in Windows, you need to use Cygwin (`https://www.cygwin.com/`) or NMake (`https://docs.microsoft.com/en-us/cpp/build/reference/nmake-reference`).

We will create the makefile to automate the steps that we have performed in this chapter. This will make it easy to do the process repetitively when required without typing it manually. We are going to create a makefile that will do tasks such as the following:

*   Bringing up/down Postgres

*   Generating code using sqlc

The makefile can be seen in the `chapter1` directory; the following shows a snippet of the script:

```
..

.PHONY : postgresup postgresdown psql createdb teardown_
recreate generate
```

```
postgresup:
    docker run --name test-postgres -v $(PWD):/usr/share/
chapter1 -e POSTGRES_PASSWORD=$(DB_PWD) -p 5432:5432 -d $(DB_
NAME)
```

...

```
# task to create database without typing it manually
createdb:
    docker exec -it test-postgres psql $(PSQLURL) -c "\i /usr/
share/chapter1/db/schema.sql"
```

...

With the makefile, you can now bring up the database easily using this command:

```
make postgresup
```

The following is used to bring down the database:

```
make postgresdown
```

sqlc will need to be invoked to regenerate the auto-generated code whenever changes are made to the schema and SQL queries. You can use the following command to regenerate the files:

```
make generate
```

Summary

In this chapter, we have covered the different stages that we need to go through to set up the database for our fitness application. We have also written a makefile to save us time by automating different database-related tasks that will be needed for the development process.

In the next chapter, we will look at logging for our sample application. Logging is a simple, yet crucial component. Applications use logging to provide visibility into the running state of an application.

2

Application Logging

Building any kind of application to fulfill a user's need is one piece of the puzzle; another piece is figuring out how we are going to design it so that we can support it in case there are issues in production. Logging is one of the most important things that need to be thought about thoroughly to allow some visibility when a problem arises. Application logging is the process of saving application events and errors; put simply, it produces a file that contains information about events that occur in your software application. Supporting applications in production requires a quick turnaround, and to achieve this, sufficient information should be logged by the application.

In this chapter, we will look at building a logging server that will be used to log events (e.g., errors) from our application. We will also learn how to multiplex logging to allow us to log different events based on how we configure it. We will cover the following in this chapter:

- Exploring Go standard logging
- Local logging
- Writing log messages to the logging server
- Configuring multiple outputs

Technical requirements

All the source code explained in this chapter can be checked out at `https://github.com/ PacktPublishing/Full-Stack-Web-Development-with-Go/tree/main/Chapter02`, while the logging server can be checked out at `https://github.com/PacktPublishing/ Full-Stack-Web-Development-with-Go/tree/main/logserver`

Exploring Go standard logging

In this section, we will look at the default logging library provided by the Go language. Go provides a rich set of libraries; however, like every other library, there are limitations – it does not provide leveled logging (`INFO`, `DEBUG`, etc.), file log file features, and many more. These limitations can be overcome by using open source logging libraries.

Go provides very diverse and rich standard libraries for applications. Logging is one of them, and it is available inside the `log` package. The following documentation link provides complete information on the different functions available inside the `https://pkg.go.dev/log@latest` package.

Another package that is available in Go standard library is the `fmt` package, which provides functions for I/O operations such as printing, input, and so on. More information can be found at `https://pkg.go.dev/fmt@latest`. The available functions inside the `log` package are similar to the `fmt` package, and when going through the sample code, we will see that it is super easy to use.

The following are some of the functions provided by the `log` package (`https://pkg.go.dev/log`):

```
func (1 *Logger) Fatal(v ...interface{})
func (1 *Logger) Fatalf(format string, v ...interface{})
func (1 *Logger) Fatalln(v ...interface{})


func (1 *Logger) Panic(v ...interface{})


func (1 *Logger) Prefix() string
func (1 *Logger) Print(v ...interface{})
func (1 *Logger) Printf(format string, v ...interface{})
func (1 *Logger) Println(v ...interface{})
func (1 *Logger) SetFlags(flag int)
func (1 *Logger) SetOutput(w io.Writer)
func (1 *Logger) SetPrefix(prefix string)
```

Let's take a look at the example code from the sample repository, `https://github.com/PacktPublishing/Full-Stack-Web-Development-with-Go/tree/main/Chapter02`. The `main.go` file resides inside `example/stdlog`. To understand how to use the `log` package, build and run the code:

```
go run .
```

On a successful run, you will get the following output:

```
2021/10/15 10:12:38 Just a log text
main.go:38: This is number 1
10:12:38 {
        «name»: «Cake»,
        «batters»: {
```

```
                «batter»: [
                        {
                                «id»: «001»,
                                «type»: «Good Food»
                        }
                ]
        },
        «topping»: [
                {
                        «id»: «002»,
                        «type»: «Syrup»
                }
        ]
}
```

The output shows that the standard logging library is configurable to allow different log output formats – for example, you can see in the following that the message is prefixed with the formatted date and time:

```
2021/10/15 10:12:38 Just a log text
```

The function that takes care of formatting the prefix for logging is the `SetFlags(..)` function:

```
func main() {
  ...
  // set log format to - dd/mm/yy hh:mm:ss
  ol.SetFlags(log.LstdFlags)
  ol.Println(«Just a log text»)
  ...
}
```

The code sets the flag to use `LstdFlags`, which is a combination of date and time. The following table shows the different flags that can be used:

Flag	Explanation
Ldate	A flag to specify the date in the local time zone in the format YYYY/MM/DD
Ltime	A flag to specify time using the local time zone in the format HH:MM:SS
Lmicroseconds	A flag to specify in microseconds
Llongfile	A flag to specify the filename and line number

Lshortfile	The final filename element and line number
LUTC	When using the Ldate or Ltime flag, we can use this flag to specify using UTC instead of the local time zone
Lmsgprefix	A flag to specify the prefix text to be shown before the message
LstdFlags	This flag uses the standard flag that has been defined, which is basically Ldate or Ltime

The standard library can cover some use cases for application log requirements, but there are times when applications require more features that are not available from the standard library – for example, sending log information to multiple outputs will require extra functionality to be built, or in another scenario, you might need to convert nested error logs into JSON format. In the next section, we will explore another alternative for our sample application.

Using golog

Now that we understand what is available in the standard library, we want to explore the option of using a library that can provide us with more flexibility. We will look at the golog open source project (https://github.com/kataras/golog). The golog library is a dependency-free logging library that provides functionality such as leveled logging (INFO, ERROR, etc.), JSON-based output, and configurable color output.

One of the most used features of logging is log levels, also known as leveled logging. Log levels are used to categorize output information from an application into different severity levels. The following table shows the different severity levels:

INFO	Just for information purposes
WARN	Something is not running correctly, so keep an eye out for it in case there are more severe errors
ERROR	There is an error that will need some attention
DEBUG	Information that is important to assist in troubleshooting in production, or added into the application for tracing purposes
FATAL	Something bad happened in the application that requires immediate response/investigation

Example code can be found inside the example/golog directory. Build and run the code, and you will get the following output:

```
2021/10/15 13:43 This is a raw message, no levels, no colors.
[INFO] 2021/10/15 13:43 This is an info message, with colors (if the output is terminal)
[WARN] 2021/10/15 13:43 This is a warning message
[ERRO] 2021/10/15 13:43 This is an error message
[DBUG] 2021/10/15 13:43 This is a debug message
[FTAL] 2021/10/15 13:43 Fatal will exit no matter what
```

Figure 2.1 – Example of golog output

Each prefix of the log messages is of a different color, which corresponds to the different severity levels; this is useful when you are going through a long list of log messages. Different log levels are assigned different colors to make it easy to go through them.

The code to generate this log is similar to the standard library code, as shown here:

```
func main() {
  golog.SetLevel(«error»)

  golog.Println(«This is a raw message, no levels, no
                colors.»)
  golog.Info(«This is an info message, with colors (if the
            output is terminal)»)
  golog.Warn(«This is a warning message»)
  golog.Error(«This is an error message»)
  golog.Debug(«This is a debug message»)
  golog.Fatal(`Fatal will exit no matter what,
            but it will also print the log message if
            logger›s Level is >=FatalLevel`)
}
```

The library provides level-based logging. This means that the library can show log messages based on what is configured to be shown; for example, for development, we want to configure the logger to show all log messages, while in production, we want to show only error messages. The following table shows what the output will look like when different levels are configured for golog:

Level	Output
`golog.SetLevel("info")`	2021/10/15 12:07 This is a raw message, no levels, no colors. [INFO] 2021/10/15 12:07 This is an info message, with colors (if the output is terminal) [WARN] 2021/10/15 12:07 This is a warning message [ERRO] 2021/10/15 12:07 This is an error message [FTAL] 2021/10/15 12:07 Fatal will exit no matter what
`golog.SetLevel("debug")`	2021/10/15 12:08 This is a raw message, no levels, no colors. [INFO] 2021/10/15 12:08 This is an info message, with colors (if the output is terminal) [WARN] 2021/10/15 12:08 This is a warning message [ERRO] 2021/10/15 12:08 This is an error message [DBUG] 2021/10/15 12:08 This is a debug message [FTAL] 2021/10/15 12:08 Fatal will exit no matter what
`golog.SetLevel("warn")`	2021/10/15 12:08 This is a raw message, no levels, no colors. [WARN] 2021/10/15 12:08 This is a warning message [ERRO] 2021/10/15 12:08 This is an error message [FTAL] 2021/10/15 12:08 Fatal will exit no matter what

golog.SetLevel("error")	2021/10/15 12:11 This is a raw message, no levels, no colors.
	[ERRO] 2021/10/15 12:11 This is an error message
	[FTAL] 2021/10/15 12:11 Fatal will exit no matter what
golog.SetLevel("fatal")	2021/10/15 12:11 This is a raw message, no levels, no colors.
	[FTAL] 2021/10/15 12:11 Fatal will exit no matter what

We covered golog and its features in this section, and now we have a good understanding of the different options available for us to use for logging. In the next section, we will look at golog a bit more.

Local logging

Now that we have an idea of how to use golog, we are going to use more of its features to extend it. The library provides a function allowing applications to handle writing the log messages for each log level – for example, an application wants to write all errors into a file while the rest print out into the console.

We are going to take a look at the example code inside the example/gologmoutput directory. Build and run it and you will see two new files created called infoerr.txt and infolog.txt. The output from both files will look as follows:

```
[ERRO] 2021/11/26 21:11 This is an error message [INFO]
2021/11/26 21:11 This is an info message, with colors (if the
output is terminal)
```

The app uses the os.OpenFile standard library to create or append files called infolog.txt and infoerr.txt, which will contain different log information that is configured using the golog SetLevelOutput function. The following is the snippet of the function that configured the different logging output using golog:

```go
func configureLogger() {
    // open infolog.txt  append if exist (os.O_APPEND) or
    // create if not (os.O_CREATE) and read write
    // (os.O_WRONLY)
    infof, err := os.OpenFile(logFile,
                os.O_APPEND|os.O_CREATE|os.O_WRONLY,
                0666)
```

```
   ...
   golog.SetLevelOutput(«info», infof)

    // open infoerr.txt  append if exist (os.O_APPEND) or
        create if not (os.O_CREATE) and read write
    // (os.O_WRONLY)
    // errf, err := os.OpenFile(«infoerr.txt»,
                os.O_APPEND|os.O_CREATE|os.O_WRONLY,
                0666)

    ...
   golog.SetLevelOutput(«error», errf)
}
```

The rest of the log-level messages are written to stdout, which is configured by default by the library.

In this section, we learned how to configure golog to allow us to log errors and information separately. This is super useful as, in production, we will have a hard time if we log everything into a single file. In the next section, we will look at building our own simple logging server to accept log requests from our application.

Writing log messages to the logging server

In the modern cloud environment, there are multiple instances of the same application running on different servers. Due to the distributed nature of the cloud environment, it will be hard to keep track of the different logs produced by the different application instances. This will require using a centralized logging system that will be able to capture all the different log messages from the different applications and systems.

For our needs, we will build our own logging server to capture all log messages in one single place; the code can be found at https://github.com/PacktPublishing/Full-Stack-Web-Development-with-Go/tree/main/logserver. The logging server will be a central place that will collate log information from our application, which will help in troubleshooting when our applications are deployed in a cloud environment. The downside of having a central logging server is that when the logging server goes down, we have no visibility of the logging information except by going to the server that hosts the applications.

REST stands for **representational state transfer**; in layman's terms, it describes a server that uses the HTTP protocol and methods to communicate to resources in the server. Information is delivered in different formats, with JSON being the most popular format. It is language agnostic, which means that the logging server can be used by any application that can send and receive over HTTP.

On a successful build, the logging server will display the following message:

```
2021/10/15 23:37:31 Initializing logging server at port 8010
```

Once the logging server is up, go back to the chapter2 root directory where the sample app resides and test the app by running the following command:

```
make build
```

On completion, run the new binary called sampledb. The sampledb app will send log messages to the logging server:

```
"{\n  \"timestamp\": 1634301479,\n  \"level\":
\"info\",\n  \"message\": \"Starting the application...\"\n}\n"
"{\n  \"timestamp\": 1634301479,\n  \"level\":
\"info\",\n  \"message\": \"Database connection fine\"\n}\n"
"{\n  \"timestamp\": 1634301479,\n  \"level\":
\"info\",\n  \"message\": \"Success - user creation\"\n}\n"
"{\n  \"timestamp\": 1634301479,\n  \"level\":
\"info\",\n  \"message\": \"Success - exercise creation\"\n}\n"
"{\n  \"timestamp\": 1634301479,\n  \"level\":
\"info\",\n  \"message\": \"Application complete\"\n}\n"
"{\n  \"timestamp\": 1634301479,\n  \"level\":
\"info\",\n  \"message\": \"Application complete\"\n}\nut\"\
n}\n"
```

The logging server runs as a normal HTTP server that listens on port 8010, registering a single endpoint, /log, to accept incoming log messages. Let's go through it and try to understand how the logging server works. But before that, let's take a look at how the server code works:

```
import (
  ...
  «github.com/gorilla/mux»
  ...
)


func runServer(addr string) {
  router = mux.NewRouter()
  initializeRoutes()

  ...
```

```
    log.Fatal(http.ListenAndServe(addr, router))
}
```

The server is using a framework called Gorilla Mux (`github.com/gorilla/mux`), which is responsible for accepting and dispatching incoming requests to their respective handler. The `gorilla/mux` package that we are using for this sample is used actively by the open source community; however, it is, at the moment, looking for a maintainer to continue the project.

The handler that takes care of handling the request is inside `initializeRoutes()`, as shown here:

```
func initializeRoutes() {
  router.HandleFunc(«/log», loghandler).Methods(http.
    MethodPost)


}
```

The `router.HandleFunc(..)` function configured the `/log` endpoint, which will be handled by the `loghandler` function. `Methods("POST")` is instructing the framework that it should accept only the POST HTTP method for incoming requests that hit the `/log` endpoint.

Now we are going to take a look at the `loghandler` function, which is responsible for processing the incoming log messages:

```
func loghandler(w http.ResponseWriter, r *http.Request) {
  body, err := ioutil.ReadAll(r.Body)


  ...

  w.WriteHeader(http.StatusCreated)
}
```

The `http.ResponseWriter` parameter is a type that is defined as an interface to be used to construct an HTTP response – for example, it contains the `WriteHeader` method, which allows writing header into the response. The `http.Request` parameter provides an interface for the function to interact with the request received by the server – for example, it provides a `Referer` function to obtain a referring URL.

The `loghandler` function does the following:

1. Reads the request body as it contains the log message.

2. On successful reading of the body, the handler will return HTTP status code 201 (StatusCreated). Code 201 means the request has been processed successfully and the resource (in this case, the log JSON message) has been created successfully, or in our case, printed successfully.

3. Prints out the log message to stdout.

For more detailed information about the different standard HTTP status codes, refer to the following website: https://developer.mozilla.org/en-US/docs/Web/HTTP/Status.

We have learned how to add logs to an application and how to build a simple logging server that can be hosted separately from our application. In the next section, we will create a logging wrapper that will allow our application to choose whether it wants to log locally or log to a server.

Configuring multiple outputs

Why do we want to configure multiple outputs? Well, it is useful as, during development, it is easier to look at logs locally for troubleshooting purposes, but in production, it's not possible to look at a log file, as everything will be inside the logging server.

We are going to write a thin layer of wrapper code that will wrap the golog library; the code that we are going to look at is inside the chapter2/ directory, inside the logger/log.go file. The benefit of having a wrapper code for the golog library is to isolate the application for interfacing directly with the library; this will make it easy to swap to different logging libraries when and if required. The app configured the wrapper code by passing the parsed flag to the SetLoggingOutput(..) function.

Build the application by running the following:

```
make build
```

Then, run it, passing the flag to true as follows to write the log message to stdout:

```
./sampledb -local=true
```

The debug log will be printed out in stdout:

```
[DBUG] 2021/10/17 19:15 Application logging to stdout = true
```

Figure 2.2 – Log output from sampledb

All info log messages will be printed out into the `logs.txt` file:

```
[INFO] 2021/10/17 19:16 Starting the application...
[INFO] 2021/10/17 19:16 Database connection fine
[INFO] 2021/10/17 19:16 Success - user creation
[INFO] 2021/10/17 19:16 Success - exercise creation
[INFO] 2021/10/17 19:16 Success - updating workout
[INFO] 2021/10/17 19:16 Application complete
```

Figure 2.3 – Log messages inside logs.txt

The logger is configured by the application using the `local` flag by calling the `SetLoggingOutput(..)` function:

```
func main() {
    l := flag.Bool(«local», false, «true - send to stdout, false
                    - send to logging server»)
    flag.Parse()

    logger.SetLoggingOutput(*l)

    logger.Logger.Debugf(«Application logging to stdout =
                    %v», *l)
    ...
```

Two main functions in the wrapper code do most of the wrapping of the `golog` framework:

- `configureLocal()`
- `configureRemote()`

```
...
func configureLocal() {
    file, err := os.OpenFile(«logs.txt»,
                os.O_APPEND|os.O_CREATE|os.O_WRONLY, 0666)
    ...
    Logger.SetOutput(os.Stdout)
    Logger.SetLevel(«debug»)
    Logger.SetLevelOutput(«info», file)
}
...
```

The `configureLocal()` function is responsible for configuring logging to write to both `stdout` and the configured file named `logs.txt`. The function configured golog to set the output to `stdout` and the level to `debug`, which means that everything will be going to `stdout`.

The other function is `configureRemote()`, which configures golog to send all messages to the remote server in JSON format. The `SetLevelOutput()` function accepts the `io.Writer` interface, which the sample app has implemented to send all info log messages:

```
//configureRemote for remote logger configuration
func configureRemote() {
  r := remote{}
  Logger.SetLevelFormat(«info», «json»)
  Logger.SetLevelOutput(«info», r)
```

The `Write(data []byte)` function performs a POST operation, passing the log message to the logging server:

```
func (r remote) Write(data []byte) (n int, err error) {
  go func() {
    req, err := http.NewRequest("POST",
      «http://localhost:8010/log»,
      bytes.NewBuffer(data),
    )

    ...
      resp, _ := client.Do(req)
      defer resp.Body.Close()
    }
  }()
  return len(data), nil
}
```

In this final section, we have learned how to create configurable logging that will allow applications to log either locally or remotely. This helps our application to be prepared and deployable in different environments.

Summary

In this chapter, we have looked at different ways of adding log functionality to applications. We also learned about the `golog` library, which provides more flexibility and features than the standard library can offer. We looked at creating our own simple logging server that enables our application to send log information that can be used in a multi-service environment.

In the next chapter, we will look at how to add observability functionality to applications. We will look at tracing and metrics and go through the OpenTelemetry specification.

3

Application Metrics and Tracing

In *Chapter 2*, *Application Logging*, we looked at logging, and how we use logging inside our backend Go code. In this chapter, we will proceed to look at monitoring and tracing. To monitor and trace the application, we will look into different open source tools and libraries.

We have started building our application, and now we need to start looking into how we are going to support it. Once an application is running in production, we need to see what's happening in the application. Having this kind of visibility will allow us to understand problems that come up. In software systems, we will often come across the concept of *observability*. The concept refers to the ability of software systems to capture and store data used for analysis and troubleshooting purposes. This includes the processes and tools used in order to achieve the goal of allowing users to observe what's happening in the system.

In this chapter, we'll be covering the following topics:

- Understanding the OpenTelemetry specification

- Tracing applications

- Adding metrics to our application using Prometheus

- Running `docker-compose`

Technical requirements

All the source code explained in this chapter is available from GitHub here: `https://github.com/PacktPublishing/Full-Stack-Web-Development-with-Go/tree/main/Chapter03`.

We will be using another tool called OpenTelemetry, which will be explained in the next section, and the version that we use in this book is v1.2.0, available here: `https://github.com/open-telemetry/opentelemetry-go/tree/v1.2.0`.

Understanding OpenTelemetry

OpenTelemetry is an open source project that enables developers to provide observability capability to their applications. The project provides a Software Development Kit (SDK) for different programming languages, with Go as one of the supported languages, which is integrated with the application. The SDK is for metric collection and reporting, as it provides integration with different open source frameworks, making the integration process seamless. OpenTelemetry also provides a common standard, providing the application flexibility to report the collected data to different observability backend systems. OpenTelemetry's website is at `https://opentelemetry.io/`.

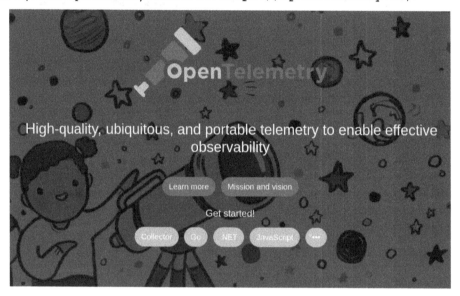

Figure 3.1 – OpenTelemetry logo

OpenTelemetry is actually the merging of the OpenTracing and OpenCensus projects. The project is used to instrument, collect, and export metrics, logs, and traces. OpenTelemetry can be used across several languages, and Go is one of the supported languages.

The main benefit of following the OpenTelemetry specification is that it is vendor-agnostic, which means that applications written using their APIs are portable across different observability vendors. For example, applications that are written to write metrics into a filesystem will require a few lines of code changes to allow it to store metrics in Prometheus, which we will discuss in the *Adding metrics using Prometheus* section.

The two main components of OpenTelemetry are the following:

- **Tracing**: This provides applications with the capability to track service requests as they flow through systems by collecting data. For example, with the tracing capability, we can see how an HTTP request flows through the different systems in the network.

- **Metrics**: This provides applications with the ability to collect and store measurements for detecting performance anomalies and forecasting. For example, collecting metrics in our application will give us visibility into how long a database query takes or how long it takes to process a certain batch job.

You can find the OpenTelemetry specification at the following link: `https://opentelemetry.io/docs/reference/specification/`.

The specification allows users to plug-and-play different OpenTelemetry implementations easily without any dependency on single-vendor libraries. This means that all the relevant contracts that are outlined in the specification document can be implemented. Some concepts are important to understand in order to use OpenTelemetry effectively. The following are the concepts that are relevant to the specification:

- **Components**: These are basically the core vendor-agnostic specifications, outlining the different parts of the system that need to be implemented. The components are collectors, the APIs, the SDK, and instrumenting libraries.

- **Data sources**: This is the data that the specification supports: traces, logs, metrics, and baggage.

- **Instrumenting and libraries**: There are two ways to integrate the provided library – either automatically by using the library provided by the vendor or open source contribution, or manually as per the application requirements.

In the next section, we are going to look at the implementation side of the specification, which involves both the APIs and the SDK.

The OpenTelemetry APIs and SDK

OpenTelemetry is made of several components, and two of the main components that we are going to talk about are the APIs and SDK. The specification defines cross-language requirements that any implementation must adhere to as part of the requirements:

- The **APIs**: This defines the data types and operations that will be used to generate telemetry data

- The **SDK**: This defines the implementation of the APIs for processing and exporting capabilities

There is a clear distinction between the APIs and SDK – it's clear that the APIs are contracts that are provided by the specification, while the SDK provides the different functionalities required to allow metrics data to be processed and exported. Metrics data contains information such as memory used, CPU usage, etc.

The specification provides an API for the following:

- **Context**: This contains the values that are carried around across API calls. This is data that can be passed between system calls and carry application information.

- **Baggage**: A set of name-value pairs describing user-defined properties.
- **Tracing**: An API definition that provides the tracing functionality
- **Metrics**: An API definition that provides the metric recording functionality

We will look at how the OpenTelemetry tracing API looks and how to add the tracing capability to applications.

Tracing applications

In the previous chapter, we learned about logging and how logging can give us visibility into what's going on inside our application. The line between logging and tracing is blurry; what we need to understand is that logging just provides information on what a process is currently doing, while tracing gives us cross-cutting visibility across different components, allowing us to get a better understanding of the data flow and time taken for a process to complete.

For example, with tracing, we can answer questions such as the following:

- How long does the add-to-cart process take?
- How long does it take to download a payment file?

We will go through the different APIs that are outlined in the specification and implement those APIs using the implementation provided by the OpenTelemetry library.

The following figure shows the links between different entities.

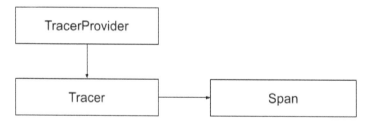

Figure 3.2 – Tracing an API relationship

TracerProvider is the entry point to use the tracing API and it provides access to **Tracer**, which is responsible for creating **Span**. **Span** is used to trace an operation in our application. Before we move further to the next layer, which is the SDK, we will take a look briefly at **Jaeger**, which is one of the support tools provided by the OpenTelemetry library for tracing.

Installing Jaeger

Jaeger (`https://www.jaegertracing.io/`) is a popular open source distributed tracing platform; it provides its own client libraries for a wide variety of programming languages, which can be seen at `https://github.com/orgs/jaegertracing/repositories`. We will be running Jaeger as a Docker container to reduce the amount of setup that is required when installing the application manually. Let's start up Jaeger using the following `docker` command:

```
docker run --name jaeger \
  -p 5775:5775/udp \
  -p 6831:6831/udp \
  -p 6832:6832/udp \
  -p 5778:5778 \
  -p 16686:16686 \
  -p 14268:14268 \
  -p 14250:14250 \
  -p 9411:9411 \
  jaegertracing/all-in-one:latest
```

On successful launch, there will be a lot of logs printed that look like the following:

```
{"level":"info","ts":1637930923.8576558,"caller":"flags/
service.go:117","msg":"Mounting metrics handler on admin
server","route":"/metrics"}
{"level":"info","ts":1637930923.857689,"caller":"flags/
service.go:123","msg":"Mounting expvar handler on admin
server","route":"/debug/vars"}
{"level":"info","ts":1637930923.8579082,"caller":"flags/
admin.go:104","msg":"Mounting health check on admin
server","route":"/"}
{"level":"info","ts":1637930923.8579528,"caller":"flags/
admin.go:115","msg":"Starting admin HTTP server","http-
addr":":14269"}
...

...
{"level":"info","ts":1637930923.8850179,"caller":"app/
server.go:258","msg":"Starting HTTP server","port":16686,"a
ddr":":16686"}
{"level":"info","ts":1637930923.8850145,"caller":"healthcheck/
handler.go:129","msg":"Health Check state
change","status":"ready"}
```

```
{"level":"info","ts":1637930923.8850334,"caller":"app/
server.go:277","msg":"Starting GRPC server","port":16685,"a
ddr":":16685"}
{"level":"info","ts":1637930924.8854718,"caller":"channelz/
logging.go:50","msg":"[core]Subchannel Connectivity change to
IDLE","system":"grpc","grpc_log":true}
{"level":"info","ts":1637930924.8855824,"caller":"grpclog/
component.go:71","msg":"[core]pickfirstBalancer:
UpdateSubConnState: 0xc00003af30, {IDLE connection error: desc
= \"transport: Error while dialing dial tcp :16685: connect:
connection refused\"}","system":"grpc","grpc_log":true}
{"level":"info","ts":1637930924.885613,"caller":"channelz/
logging.go:50","msg":"[core]Channel Connectivity change to
IDLE","system":"grpc","grpc_log":true}
```

Jaeger is now ready, the tool is not a desktop application but it provides a user interface that is accessible using the browser. Open your browser and type in the following URL: `http://localhost:16686`. It will open the Jaeger main page (*Figure 3.3*):

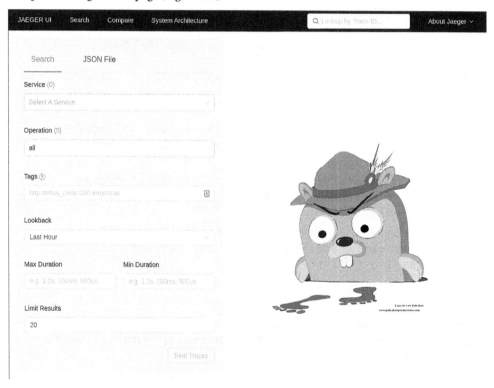

Figure 3.3 – Jaeger main page

At the moment, Jaeger does not contain anything, as there are no applications that are using it.

Integrating the Jaeger SDK

Now that Jaeger is ready, let's look at how we are going to write tracing information using OpenTelemetry. The library provides support for the Jaeger SDK out of the box; this allows applications to use the API to write tracing to Jaeger.

The example that we will be using in this section is inside the jaeger/opentelem/trace directory in the chapter's GitHub repository. The file that we want to look at is tracing.go as shown here:

```go
package trace

import (
        «context»

        «go.opentelemetry.io/otel"

        «go.opentelemetry.io/otel/exporters/jaeger"
        «go.opentelemetry.io/otel/sdk/resource"
        «go.opentelemetry.io/otel/sdk/trace"
        sc "go.opentelemetry.io/otel/semconv/v1.4.0"
)

type ShutdownTracing func(ctx context.Context) error

func InitTracing(service string) (ShutdownTracing, error)
{
  // Create the Jaeger exporter.
  exp, err := jaeger.New(jaeger.WithCollectorEndpoint())
  if err != nil {
   return func(ctx context.Context) error { return nil },
     err
  }

  // Create the TracerProvider.
  tp := trace.NewTracerProvider(
          trace.WithBatcher(exp),
```

```
            trace.WithResource(resource.NewWithAttributes(
                    sc.SchemaURL,
                    sc.ServiceNameKey.String(service),
            )),
    )

    otel.SetTracerProvider(tp)
    return tp.Shutdown, nil
}
```

Let's take a look at what each part of the code is doing. Line 18 is initializing the Jaeger SDK inside the OpenTelemetry library. On successfully initializing the Jaeger SDK, the code continues to provide the newly created Jaeger and uses it with the OpenTelemetry library to create a new `TracerProvider` API. As discussed in the previous section, `TracerProvider` is the API that is used as the main entry for OpenTelemetry. This is performed on lines 24-30.

On obtaining `TracerProvider`, we will need to call the global `SetTracerProvider` to let OpenTelemetry know about it, which is done on line 32. Once the Jaeger SDK has been successfully initialized, now it's a matter of using it in the application.

Let's take a look at the code sample for using the tracing functionality. The sample application that we are going to look at can be found inside the `jaeger/opentelem` directory inside `main.go`.

Integration with Jaeger

We are going to go through section by section to explain what the code is doing. The following code section shows the `InitTracing` function that takes care the initialization process being called:

```
package main

import (
        t "chapter.3/trace/trace"
        "context"
        "fmt"
        "go.opentelemetry.io/otel"
        "go.opentelemetry.io/otel/attribute"
        "go.opentelemetry.io/otel/trace"
        "log"
```

```go
            "sync"
            "time"
)

const serviceName = "tracing"

func main() {
  sTracing, err := t.InitTracing(serviceName)
  if err != nil {
    log.Fatalf("Failed to setup tracing: %v\n", err)
  }
  defer func() {
    if err := sTracing(context.Background()); err != nil
    {
      log.Printf("Failed to shutdown tracing: %v\n", err)
    }
  }()
  ctx, span := otel.Tracer(serviceName)
               .Start(context.Background(), "outside")
  defer span.End()

  var wg sync.WaitGroup

  wg.Add(1)
  go func() {
    _, s := otel.Tracer(serviceName).Start(ctx, "inside")
    ...
    wg.Done()
  }()

    wg.Add(1)
    go func() {
      _, ss := otel.Tracer(serviceName).Start(ctx,
                                           "inside")

      ...
      wg.Done()
```

```
    } ()

    wg.Wait ()
    fmt.Println ("\nDone!")
}
```

Once the SDK completes the initialization process, the code can start using the API to write tracing information and this is done by getting a Span using the Tracer API as shown on lines 27-29. The code uses sync.WaitGroup (lines 35 and 45) to ensure that the main thread does not finish before the goroutine completes – the goroutine is added to simulate some kind of processing to be done to generate a trace that will be reported to Jaeger.

The Tracer API only has one Start function, which is called to initiate the tracing operation, and the tracing operation is considered complete when the End function is called on Span – so, what is Span? Span is an API for tracing an operation; it has the following interface declaration:

```
type Span interface {
    End (options ...SpanEndOption)
    AddEvent (name string, options ...EventOption)
    IsRecording () bool
    RecordError (err error, options ...EventOption)
    SpanContext () SpanContext
    SetStatus (code codes.Code, description string)
    SetName (name string)
    SetAttributes (kv ...attribute.KeyValue)
    TracerProvider () TracerProvider
}
```

Multiple spans are pieced together to create a trace; it can be thought of as a **Directed Acyclic Graph (DAG)** of spans.

> **DAGs**
>
> A DAG is a term used in mathematics and computer science. It is a graph that shows dependencies, which, in our case, are the dependencies of application traces.

Figure 3.4 shows what the composition of the trace looks like:

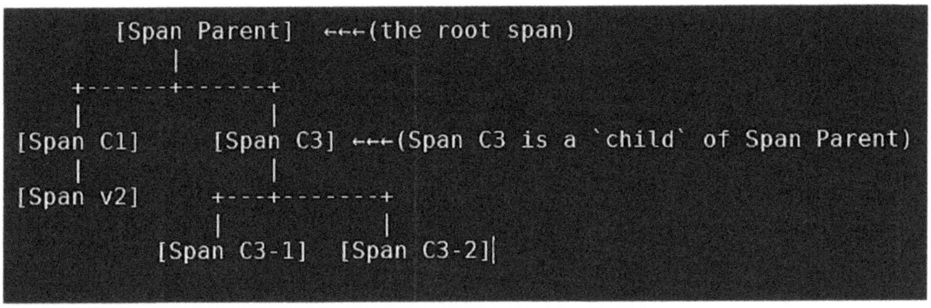

```
        [Span Parent]   ←←←(the root span)
             |
    +--------+------+
    |               |
[Span C1]       [Span C3] ←←←(Span C3 is a `child` of Span Parent)
    |               |
[Span v2]       +---+--------+
                |            |
           [Span C3-1]   [Span C3-2]|
```

Figure 3.4 – A DAG of a simple trace

The sample code creates two goroutines to perform a `sleep` operation and write trace information as shown below:

```
go func() {
    _, s := otel.Tracer(serviceName).Start(ctx, "inside")
    defer s.End()
    time.Sleep(1 * time.Second)
    s.SetAttributes(attribute.String("sleep", "done"))
    s.SetAttributes(attribute.String("go func", "1"))
    wg.Done()
}()
...
...
go func() {
    _, ss := otel.Tracer(serviceName).Start(ctx, "inside")
    defer ss.End()
    time.Sleep(2 * time.Second)
    ss.SetAttributes(attribute.String("sleep", "done"))
    ss.SetAttributes(attribute.String("go func", "2"))
    wg.Done()
}()
```

Run the complete sample application in `main.go` inside the `jaeger/opentelem` directory using the following command:

```
go run main.go
```

Upon completion, the application will write tracing information into Jaeger. Open Jaeger by accessing `http://localhost:16686` in your browser. Once it's opened, you will see a new entry under the **Service** dropdown as shown in *Figure 3.5*:

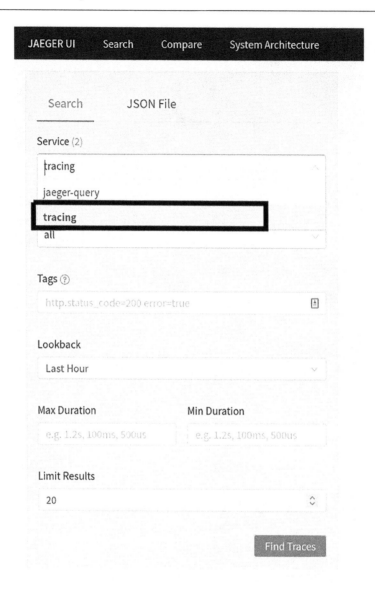

Figure 3.5 – Application trace search

The sample application tracing information is registered with the same string defined in the code, which is called `tracing`:

```
const serviceName = "tracing"
```

Clicking on the **Find Traces** button will read the trace information that is stored (*Figure 3.6*):

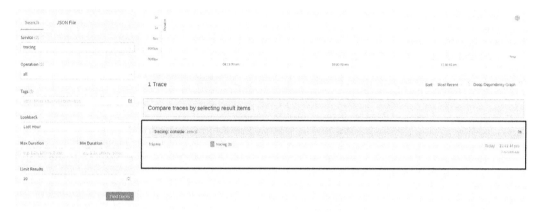

Figure 3.6 – Application traces

As can be seen in *Figure 3.6*, there is only one entry and if you click on it, it will expand more information that the app has submitted via the Span API.

Figure 3.7 – Tracing information

Figure 3.7 shows the complete tracing information, which is a composition of spans from the application. Clicking on each of the graphs will bring up more information included in the span, which is included as shown in the code here:

```
go func() {
    ...
    s.SetAttributes(attribute.String("sleep", "done"))
    s.SetAttributes(attribute.String("go func", "1"))
    ...
}()
...
go func() {
    ...
    ss.SetAttributes(attribute.String("sleep", "done"))
    ss.SetAttributes(attribute.String("go func", "2"))
```

```
    . . .
} ()
```

Now that we know how to add tracing to our application, in the next section, we will look at adding metric instrumentation that will give us visibility into some of the performance metrics relevant to our application.

Adding metrics using Prometheus

As OpenTelemetry is vendor-agnostic, it provides a wide variety of support for monitoring, exporting, and collecting metrics and one option is Prometheus. A complete list of different projects supported by OpenTelemetry can be found at `https://opentelemetry.io/registry/`. Prometheus is an open source monitoring and alerting system server that is widely used in cloud environments; it also provides libraries for a variety of programming languages.

In the previous section, we saw how to add tracing capabilities to our application and how to retrieve the traces by using Jaeger. In this section, we are going to take a look at how to create metrics using the `OpenTelemetry` library. Metrics allow us to get instrumentation information for our applications; it can provide answers to questions such as the following:

- What is the total number of requests processed in service A?
- How many total transactions are processed via payment gateway B?

Normally, collected metrics are stored for a certain amount of time to give us better insights into how the applications are performing by looking at a specific metric.

We will use the Prometheus open source project (`https://prometheus.io/`), which provides a complete monitoring solution stack and is very easy to use. The project provides a lot of features that are useful for collecting and storing metrics and monitoring our applications.

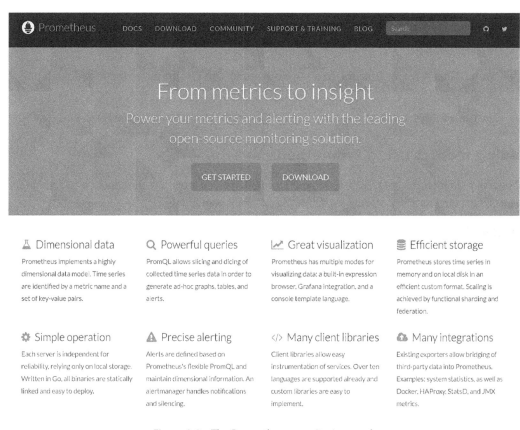

Figure 3.8 – The Prometheus monitoring stack

Similar to tracing, the OpenTelemetry specification specifies the API and SDK for metrics, as shown in *Figure 3.9*.

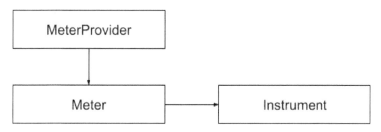

Figure 3.9 – Metrics API

The following are explanations of the metrics APIs:

- **MeterProvider**: This is an API for providing access to meters.

- **Meter**: This is responsible for creating instruments, and is unique to the instrumentation in question.

- **Instrument**: This contains the metric that we want to report; it can be synchronous or asynchronous.

Adding metrics using Prometheus

Let's start up Prometheus; make sure from your terminal that you are inside the chapter3/prom/opentelem directory and execute the following docker command:

```
docker run --name prom \
-v $PWD/config.yml:/etc/prometheus/prometheus.yml \
-p 9090:9090 prom/prometheus:latest
```

> NOTE:
> If you are using a Linux machine, use the following command:
>
> ```
> docker run --name prom \
> -v $PWD/config.yml:/etc/prometheus/prometheus.yml\
> -p 9090:9090   --add-host=host.docker.internal:host-gateway
> prom/prometheus:latest
> ```
>
> The extra parameter, --add-host=host.docker.internal:host-gateway, will allow Prometheus to access the host machine using the host.docker.internal machine name.

The config.yml file used for configuring Prometheus is inside the prom/opentelem directory and looks like the following:

```
scrape_configs:
 - job_name: 'prometheus'
   scrape_interval: 5s
   static_configs:
     - targets:
         - host.docker.internal:2112
```

We will not go through the different available Prometheus configuration options in this section. The configuration we are using informs Prometheus that we want to get metrics from the container host, which is known internally in the container as host.docker.internal, at port 2112, at an interval of 5 seconds.

Once Prometheus successfully runs, you will see the following log:

```
....
ts=2021-11-30T11:13:56.688Z caller=main.go:451 level=info fd_
limits="(soft=1048576, hard=1048576)"

...
ts=2021-11-30T11:13:56.694Z caller=main.go:996 level=info
msg="Loading configuration file" filename=/etc/prometheus/
prometheus.yml
ts=2021-11-30T11:13:56.694Z caller=main.go:1033 level=info
msg="Completed loading of configuration file" filename=/
etc/prometheus/prometheus.yml totalDuration=282.112μs db_
storage=537ns remote_storage=909ns web_handler=167ns query_
engine=888ns scrape=126.942μs scrape_sd=14.003μs notify=608ns
notify_sd=1.207μs rules=862ns
ts=2021-11-30T11:13:56.694Z caller=main.go:811 level=info
msg="Server is ready to receive web requests."
```

Next, open your browser and type in the following: `http://localhost:9090`. You will be shown the main Prometheus UI:

Figure 3.10 – The Prometheus UI

Figure 3.11 shows the way Prometheus collects metrics via a pulling mechanism where it *pulls* metric information from your application by connecting to port `2112`, which is exposed by the HTTP server running in the application. We will see later that most of the heavy lifting is done by the `OpenTelemetry` library; our application will just have to provide the metric that we want to report on.

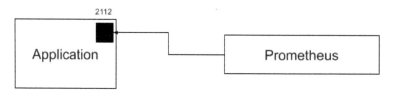

Figure 3.11 – Prometheus metric collection

Now that Prometheus is ready, we can start recording metrics to for our application. Run the application inside the `prom/opentelem` directory as follows:

```
go run main.go
```

Let the application run for a bit and you will see the following log:

```
2021/11/30 22:42:08 Starting up server on port 8000
2021/11/30 22:42:12 Reporting metric metric.random
2021/11/30 22:42:22 Reporting metric metric.random
2021/11/30 22:42:32 Reporting metric metric.random
2021/11/30 22:42:47 Reporting metric metric.random
2021/11/30 22:42:57 Reporting metric metric.random
```

- `metric.totalrequest`: This metric reports the total number of requests processed by the application; the sample application has an HTTP server running on port `8000`

- `metric.random`: This metric reports a random number

With the successful run of the sample application, we can now see the metric in the Prometheus UI. Open your browser and head to `http://localhost:9090` and type in `metric_random` and you will see something such as that shown in *Figure 3.12*; click on the **Execute** button.

Figure 3.12 – metric_random metric

Select the **Graph** tab and you will see the following figure:

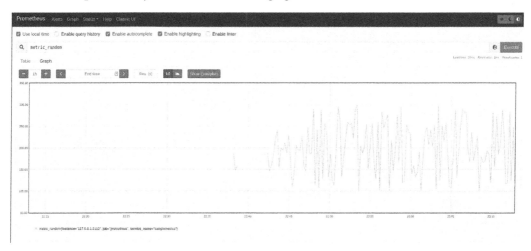

Figure 3.13 – metric_random graph

The other metric that we want to show is the total number of requests processed by the sample application's HTTP server. In order to generate some metrics, open the browser and enter `http://localhost:8000`; do so a few times so that some metrics will be generated.

Open the Prometheus UI again (`http://localhost:9090`), add the `metric_totalrequest` metric as shown in *Figure 3.14*, and click on **Execute**:

Figure 3.14 – metric_totalrequest metric

The graph will look as follows:

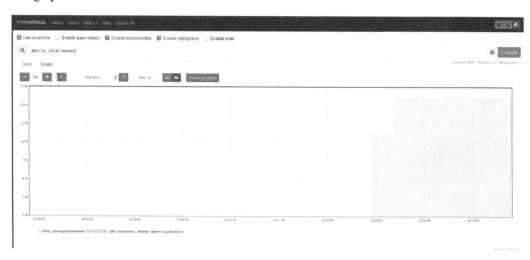

Figure 3.15 – metric_totalrequest graph

If you are having problems and cannot see the metrics, change the Prometheus configuration file, `config.yml`, inside the `chapter3/prom/opentelem` directory and change the target from `host.docker.internal` to `localhost` as shown here:

```
scrape_configs:
  - job_name: 'prometheus'
    scrape_interval: 5s
    static_configs:
        - targets:
        - localhost:2112
```

The `metrics.go` source contains the code that initializes the `otel` SDK to configure it for Prometheus, which is shown in the code snippet here:

```
package metric

...

type ShutdownMetrics func(ctx context.Context) error

// InitMetrics use Prometheus exporter
func InitMetrics(service string) (ShutdownMetrics, error) {
    config := prometheus.Config{}
```

```
    c := controller.New(
        processor.NewFactory(
            selector.NewWithExactDistribution(),
            aggregation.CumulativeTemporalitySelector(),
            processor.WithMemory(true),
        ),
        controller.WithResource(resource.NewWithAttributes(
            semconv.SchemaURL,
            semconv.ServiceNameKey.String(service),
        )),
    )
    exporter, err := prometheus.New(config, c)
    if err != nil {
      return func(ctx context.Context) error { return nil},
        err
    }

    global.SetMeterProvider(exporter.MeterProvider())

    srv := &http.Server{Addr: ":2112", Handler: exporter}
    go func() {
        _ = srv.ListenAndServe()
    }()

    return srv.Shutdown, nil
```

The following code snippet shows how it sends the metrics to Prometheus – the code can be found in main.go inside the chapter3/prom/opentelem directory:

```
package main

...
const serviceName = "samplemetrics"

func main() {
    ...
    //setup handler for rqeuest
```

```
    r.HandleFunc("/", func(rw http.ResponseWriter, r
      *http.Request) {
        log.Println("Reporting metric metric.totalrequest")

        ctx := r.Context()

        //add request metric counter
        ctr.Add(ctx, 1)


        . . .
    }).Methods("GET")


    . . .
}
```

Now that we have successfully added metrics and tracing to our applications and can view them using both Jaeger and Prometheus; in the next section, we will look at putting all the tools together to make it easy to run them as a single unit.

Running docker-compose

We normally run containers using the `docker` command, but what if we want to run more than one container in one go? This is where `docker-compose` comes to the rescue. The tool allows you to configure the different containers that you want to run as a single unit. It also allows different kinds of configurations for different containers – for example, container A can communicate via the network with container B but not with container C.

The `docker-compose` tool that we are using in this book is v2, which is the recommended version. You can find instructions for installing the tool for different operating systems here – `https://docs.docker.com/compose/install/other/`.

To make it easy to run both Prometheus and Jaeger, you can use `docker-compose`. The `docker-compose.yml` file looks as follows:

```
version: '3.3'
services:
  jaeger:
    image: jaegertracing/all-in-one:latest
    ports:
      - "6831:6831/udp"
```

```
      - "16686:16686"
      - "14268:14268"
  prometheus:
    image: prom/prometheus:latest
    volumes:
      - ./prom/opentelem/config.yml:/etc/prometheus/
        prometheus.yml
    command:
      - '--config.file=/etc/prometheus/prometheus.yml'
      - '--web.console.libraries=/usr/share/prometheus/
        console_libraries'
      - '--web.console.templates=/usr/share/prometheus/
        consoles>
    ports:
      - 9090:9090
    network_mode: "host"
```

Run docker-compose using the following command:

```
docker-compose -f docker-compose.yml   up
```

On a successful run, you will see the following log:

```
prometheus_1  | ts=2021-12-04T07:45:02.443Z caller=main.go:406
level=info msg="No time or size retention was set so using the
default time retention" duration=15d
prometheus_1  | ts=2021-12-04T07:45:02.443Z
caller=main.go:444 level=info msg="Starting
Prometheus" version="(version=2.31.1, branch=HEAD,
revision=411021ada9ab41095923b8d2df9365b632fd40c3)"
prometheus_1  | ts=2021-12-04T07:45:02.443Z caller=main.go:449
level=info build_context="(go=go1.17.3, user=root@9419c9c2d4e0,
date=20211105-20:35:02)"
prometheus_1  | ts=2021-12-04T07:45:02.443Z caller=main.go:450
level=info host_details="(Linux 5.3.0-22-generic #24+system7
6~1573659475~19.10~26b2022-Ubuntu SMP Wed Nov 13 20:0 x86_64
pop-os (none))"
prometheus_1  | ts=2021-12-04T07:45:02.444Z caller=main.go:451
level=info fd_limits="(soft=1048576, hard=1048576)"
prometheus_1  | ts=2021-12-04T07:45:02.444Z caller=main.go:452
```

```
level=info vm_limits="(soft=unlimited, hard=unlimited)"
jaeger_1        | 2021/12/04 07:45:02 maxprocs: Leaving
GOMAXPROCS=12: CPU quota undefined
prometheus_1    | ts=2021-12-04T07:45:02.445Z caller=web.go:542
level=info component=web msg="Start listening for connections"
address=0.0.0.0:9090
. . . .

. . . .

. . . .
jaeger_1        |
{"level":"info","ts":1638603902.657881,"caller":"healthcheck/
handler.go:129","msg":"Health Check state
change","status":"ready"}
jaeger_1        |
{"level":"info","ts":1638603902.657897,"caller":"app/
server.go:277","msg":"Starting GRPC server","port":16685,"a
ddr":":16685"}
jaeger_1        |
{"level":"info","ts":1638603902.6579142,"caller":"app/
server.go:258","msg":"Starting HTTP server","port":16686,"a
ddr":":16686"}
```

The up parameter we are using will start the container in the terminal and run in attached mode, which allows you to show all the logs on the screen. You also can run in detached mode to run the container in the background as follows:

```
docker-compose -f docker-compose.yml  up -d
```

Summary

In this section, we looked at how to add metrics and tracing into an application using the OpenTelemetry library. Having this observability in an application will enable us to troubleshoot issues faster and also keep track of the performance of our application from the provided metrics. We also took a look at using two different open source projects that allow us to look at the data collected from our application.

In this chapter, we looked at the plumbing and infrastructure required to monitor and trace our application. In the next chapter, we will look at different aspects of building both dynamic and static content for our web application and how to package the application to make it easier to deploy anywhere.

Part 2:
Serving Web Content

Upon completing this part of the book, you will be able to create server-rendered pages using an HTML/template and Gorilla Mux. You will also learn how to create and expose an API that will be used by the frontend. Securing the API will be discussed, including middleware.

This part includes the following chapters:

- *Chapter 4, Serving and Embedding HTML Content*
- *Chapter 5, Securing the Backend and Middleware*
- *Chapter 6, Moving to API-First*

4

Serving and Embedding HTML Content

As we build on our foundations, it is important that we look at another aspect of processing HTTP user requests, routing. Routing is useful as it allows us to structure our application to handle different functionality for certain HTTP methods, such as a GET that can retrieve and a POST on the same route that can replace the data. This concept is the fundamental principle of designing a REST-based application. We'll end the chapter by looking at how we can use the new embed directive introduced in Go version 1.16 to bundle our web app as a single self-contained executable. This chapter will provide us with the tools to handle user data and create the interface for the user.

By the end of this chapter, you will have learned how static and dynamic content is served by the application. You will also have learned how to embed all the different assets (icons, .html, .css, etc.) that will be served by the web application in the application using a single binary. In this chapter, we'll cover the following topics:

- Handling HTTP functions and Gorilla Mux
- Rendering static and dynamic content
- Using Go embed to bundle your content

Technical requirements

All the source code for this chapter can be accessed at https://github.com/PacktPublishing/Full-Stack-Web-Development-with-Go/tree/main/Chapter04.

Handling HTTP functions and Gorilla Mux

When we look at the **Go standard library**, we can see that a lot of thought has gone into the **HTTP library**. You can check out the documentation for the Go standard library here: https://pkg.go.dev/net/http. However, we'll cover the foundations and look at how we can build upon them.

It's interesting to Note that the Go standard library covers both client- and server-side implementations. We will only be focusing on the parts we require to serve content.

We will create a simple app that replies with **Hello, World**, as well as look at returning POST data once we have expanded our routes.

Hello, World with defaults

The basic concepts of creating a server in **Golang** are as follows:

```
1    package main
2
3    import (
4        "fmt"
5        "log"
6        "net/http"
7        "os"
8        "time"
9    )
10
11    func handlerGetHelloWorld(wr http.ResponseWriter,
                                 req *http.Request) {
12        fmt.Fprintf(wr, "Hello, World\n")
13        log.Println(req.Method)  // request method
14        log.Println(req.URL)     // request URL
15        log.Println(req.Header)  // request headers
16        log.Println(req.Body)    // request body)
17    }
18

 . . .

29
30    func main() {
 . . .
43        router := http.NewServeMux()
44
45        srv := http.Server{
46            Addr:           ":" + port,
47            Handler:        router,
```

```
48              ReadTimeout:    10 * time.Second,
49              WriteTimeout:   120 * time.Second,
50              MaxHeaderBytes: 1 << 20,
51      }
52
...
57      router.HandleFunc("/", handlerGetHelloWorld)
58      router.Handle("/1", dummyHandler)
59      err := srv.ListenAndServe()
60      if err != nil {
61          log.Fatalln("Couldnt ListenAndServe()",
                    err)
62      }
63  }
```

You can see this code in the Git repository under the `library-mux` sub-folder.

How this works is we define a `handlerGetHelloWorld` handler function (row 11) that is passed as a parameter to the `router.HandleFunc` function. The `HandleFunc` parameter requires a function parameter that has the following signature: `func(ResponseWriter, *Request)`.

The handler's job is to take in a request type and a `ResponseWriter` and make a decision based on the request; that is, what to write to `ResponseWriter`. In our case, the `handlerGetHelloWorld` handler will send the `Hello, World` string as a response, using the `fmt.Fprintf(...)` function. The reason why it is possible for the response to be sent back is that the `http.ResponseWriter` implements the `Write()` function, which is used inside the `fmt.Fprintf(...)` function.

We now define the following steps for the main function:

1. First, we create a router: this is what our handlers will connect to. We create our own router with `NewServeMux` (line 43). We could use the `DefaultServeMux` found in the default library, but as you will see at `https://github.com/golang/go/blob/5ec87ba554c2a83cdc188724f815e53fede91b66/src/expvar/expvar.go#L334`, it contains a few additional debugging endpoints that we may not want to expose publicly. By registering our own, we gain more control and can add the same endpoints ourselves if we want them.

2. Second, we create an instance of our server and bind it to an available port. The `Addr` field on the server specifies the address and port to bind to. In our example, we are using `9002`. Different operating systems have different restrictions on what port can be used. For example, Linux systems only allow the admin or root user to run applications that use ports between `1` and `1023`.

3. The final step is to attach our router, start the server, and get it to begin listening. This is accomplished in line 57. What we're doing here is telling the router that when it gets any HTTP request for " / ", known as the document root, it should handle the request by passing it to our handler.

4. The final function, srv.ListenAndServe() (line 59), is a blocking function that starts our server up and starts listening for incoming requests on the server's defined port. When a valid HTTP request is found, it is passed to the **mux**, which then pattern matches the route – that is, the given sequence is checked against the patterns known by the mux, and if a pattern is found for " / ", then our handler is invoked. We can run our app and visit http://localhost:9002/; we should be met with the following response from the server:

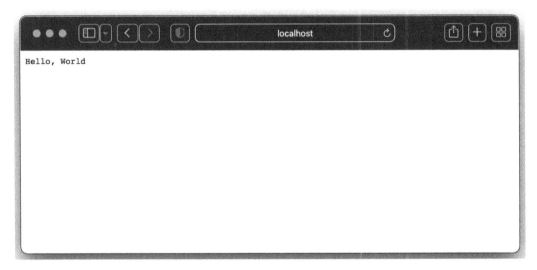

Figure 4.1 – Hello, World from Go!

It's good to note here that each request is given its own goroutine to execute concurrently, and each request's life cycle is managed by the server so we don't need to do anything explicitly to leverage this.

In the next section, we will explore building different functionalities using **Gorilla Mux**. In particular, we will look at implementing handlers and the different ways to handle HTTP methods, such as GET and POST.

Building on the basics with Gorilla Mux

Gorilla Mux, accessible at https://github.com/gorilla/mux, is a subproject of the **Gorilla project**. Gorilla Mux is an *HTTP request multiplexer* that makes it easy to match different handlers with matching incoming requests. Developers gain a lot of benefits from using the library, as it makes writing lots of boilerplate code unnecessary. The library provides advanced capabilities to match requests based on different criteria, such as schemes and dynamic URLs.

The server and router provided as part of Go's standard library are incredibly powerful for "freebies", but we're going to look at adding Gorilla Mux to our project and some of the benefits it provides.

Uses of the web consist of more than just returning *Hello World*, and generally, most web apps accept data provided by users, update the data, and even delete the data, and this is possible because the browser accepts a variety of content such as images, video, data fields, and plain text. The previous exercise focused on what is known as a GET method, which is the default sent when you load a page in your web browser, but there are many more.

The standard library implementation makes it easy to explicitly handle other types of methods, such as GET, POST, PUT, DELETE, and more, which are defined in the HTTP standard. This is typically done in the handler function as we can see below:

```
func methodFunc(wr http.ResponseWriter, req http.Request) {
    ...

    switch req.Method {
    case http.MethodGet:
        // Serve page - GET is the default when you visit a
        // site.
    case http.MethodPost:
        // Take user provided data and create a record.
    case http.MethodPut:
        // Update an existing record.
    case http.MethodDelete:
        // Remove the record.
    default:
        http.Error(wr, "Unsupported Method!",
                    http.StatusMethodNotAllowed)
    }
}
```

Let's look at an example of how we can separate two handlers, GET and POST, and some of the helpers provided by Gorilla Mux:

```
1      package main
2
3    import (
4        "bytes"
5        "fmt"
6        "io"
7        "io/ioutil"
8        "log"
9        "net/http"
10        "os"
11
12        "github.com/gorilla/mux"
13    )
14
15    func handlerSlug(wr http.ResponseWriter, req
                    *http.Request) {
16        slug := mux.Vars(req)["slug"]
17        if slug == "" {
18            log.Println("Slug not provided")
19            return
20        }
21        log.Println("Got slug", slug)
22    }
23
24    func handlerGetHelloWorld(wr http.ResponseWriter,
                        req *http.Request) {
25        fmt.Fprintf(wr, "Hello, World\n")
          // request method
26        log.Println("Request via", req.Method)
          // request URL
27        log.Println(req.URL)
          // request headers
28        log.Println(req.Header)
          // request body)
```

```
29          log.Println(req.Body)
30      }
31
32      func handlerPostEcho(wr http.ResponseWriter,
                         req *http.Request) {
            // request method
33          log.Println("Request via", req.Method)
            // request URL
34          log.Println(req.URL)
            // request headers
35          log.Println(req.Header)
36
37          // We are going to read it into a buffer
38          // as the request body is an io.ReadCloser
39          // and so we should only read it once.
40          body, err := ioutil.ReadAll(req.Body)
41
42          log.Println("read >", string(body), "<")
43
44          n, err := io.Copy(wr, bytes.NewReader(body))
45          if err != nil {
46              log.Println("Error echoing response",
                         err)
47          }
48          log.Println("Wrote back", n, "bytes")
49      }
50
51      func main() {
52          // Set some flags for easy debugging
53          log.SetFlags(log.Lshortfile | log.Ldate |
                     log.Lmicroseconds)
54
55          // Get a port from ENV var or default to 9002
56          port := "9002"
57          if value, exists :=
                os.LookupEnv("SERVER_PORT"); exists {
58              port = value
```

```
59        }
60
61        // Off the bat, we can enforce StrictSlash
62        // This is a nice helper function that means
63        // When true, if the route path is "/foo/",
          // accessing "/foo" will perform a 301
          // redirect to the former and vice versa.
64        // In other words, your application will
          // always see the path as specified in the
          // route.
65        // When false, if the route path is "/foo",
          // accessing "/foo/" will not match this
          // route and vice versa.
66
67        router := mux.NewRouter().StrictSlash(true)
68
69        srv := http.Server{
70            Addr:     ":" + port, // Addr optionally
              // specifies the listen address for the
              // server in the form of "host:port".
71            Handler: router,
72        }
73
74        router.HandleFunc("/", handlerGetHelloWorld)
              .Methods(http.MethodGet)
75        router.HandleFunc("/", handlerPostEcho)
              .Methods(http.MethodPost)
76        router.HandleFunc("/{slug}", handlerSlug)
              .Methods(http.MethodGet)
77
78        log.Println("Starting on", port)
79        err := srv.ListenAndServe()
80        if err != nil {
81            log.Fatalln("Couldnt ListenAndServe()", err)
82        }
83    }
```

We've imported the Gorilla Mux library as `mux` and set up two different handlers: `handlerGetHelloWorld` (line 24) and `handlerPostEcho` (line 32). `handlerGetHelloWorld` is the same handler we defined in the previous example that responds with *Hello, World*. Here, thanks to the extended functionality of the router, we've specified explicitly that the handler can only resolve if the user performs a `GET` method on the `"/"` endpoint (line 74).

Let's start the sample by first changing to the `chapter4/gorilla-mux` directory and running the following command:

```
go run main.go
```

We can use cURL, which is a standard utility available on Windows (use cmd instead of PowerShell) and installed by default on Linux (depending on your Linux distribution) and macOS. The tool allows users to make HTTP requests from a terminal without using a browser. Use the `curl localhost:9002` command in a separate terminal to test whether the server is up and running:

```
$ curl localhost:9002
Hello, World
$ # Specify DELETE as the option...
$ curl localhost:9002 -v -X DELETE
```

We can see that `GET` works as expected but using `-X DELETE` to tell cURL to use the `HTTP DELETE` method results in no content being returned. Under the hood, the endpoint is responding with a `405 Method Not Allowed` error message. The 405 error message reported to the user comes from the library by default.

We've added a second handler (line 75) to take data from a `POST` request. The handler for the `POST` method, `handlerPostEcho` (line 32), performs in a similar manner to the `GET` request, but we've added some additional code to read the user-provided data, store it, print it, and then return it unaltered.

We can see how this works using cURL as before:

```
$ curl -X POST localhost:9002 -d "Echo this back"
Echo this back
```

We're skipping a lot of validation and explicitly checking/handling data formats, such as JSON, at this point, but we'll build towards this in later sections.

Another benefit of using Gorilla Mux is how easy it makes pattern matching in paths. These path variables, or `slugs`, are defined using the `{name}` format or `{name:pattern}`. The following table shows different `slugs` with examples:

`/books/{pagetitle}/page/{pageno}`	`/books/mytitle/page/1`, `/books/anothertitle/page/100`
`/posts/{slug}`	/posts/titlepage
	/posts/anothertitle

Pattern can be a type of regular expression. For example, in our sample code we added a `handlerSlug` handler (line 15) to perform a simple capture. We can use cURL to test this, as shown in the following code:

```
$ curl localhost:9002/full-stack-go
...
$ # Our server will show the captured variable in its output
...
2022/01/15 14:58:36.171821 main.go:21: Got slug > full-stack-go
<
```

In this section, we have learned how to write handlers and use them with Gorilla Mux. We have also looked at configuring Gorilla Mux to handle dynamic paths that will be processed by handlers. In the next section, we will look at serving content to users from our application. The served content will contain static and dynamic content.

Rendering static content

In this section, we will learn how to serve the web pages we have created as static content. We will use the standard Go net/http package to serve up the web pages. All the code and HTML files can be found inside the static/web directory (https://github.com/PacktPublishing/Full-Stack-Web-Development-with-Go/tree/main/Chapter04/static/web).

Execute the server using the following command:

```
go run main.go
```

You will see the following message on the screen:

```
2022/01/11 22:22:03 Starting up server on port 3333 ...
```

Open your browser and enter http://localhost:3333 as the URL. You will see the login page, as shown in *Figure 4.2*:

Figure 4.2 – The login page

To access the dashboard page, you can use the URL `http://localhost:3333/dashboard.html`. You will see like the following screenshot:

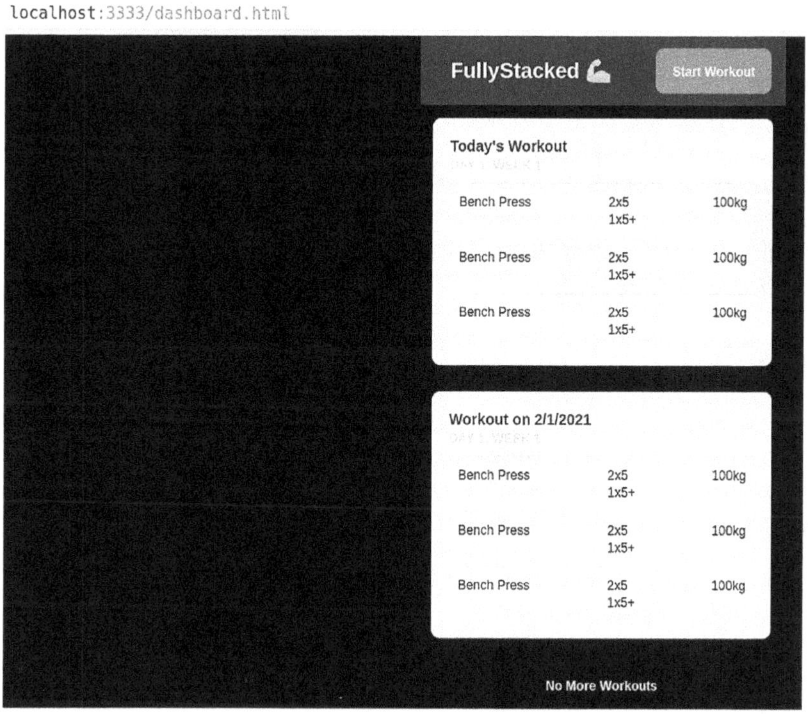

Figure 4.3 – The dashboard page

Let's take a quick look at the code that serves up the static pages:

```
1    package main
2
3    import (
4        "log"
5        "net/http"
6    )
7
8    func main() {
9        fs := http.FileServer(http.Dir("./static"))
10       http.Handle("/", fs)
11
12       log.Println("Starting up server on port 3333
                ...")
13       err := http.ListenAndServe(":3333", nil)
14       if err != nil {
15           log.Fatal("error occurred starting up
                server : ", err)
16       }
17   }
```

As can be seen, this is a simple HTTP server that uses the http.FileServer(..) Go standard library function (shown in line 9). The function is called by passing in the (./static) parameter to the directory that we want to serve (line 9). The example code can be found inside the chapter4/static/web/static folder.

Rendering dynamic content

Now that we understand how to serve static content using the net/http package, let's take a look at adding some dynamic content using Gorilla Mux found here: https://github.com/PacktPublishing/Full-Stack-Web-Development-with-Go/tree/main/Chapter04/dynamic. Execute the server using the following command:

```
go run main.go
```

Launch your browser and enter http://localhost:3333 as the address; you will see a login screen similar to the static content. Perform the following steps on the login screen:

1. Enter any combination of username and password on the login screen.

2.   Click the **Login** button.

You will get a **Login unsuccessful** message, as shown in *Figure 4.4*.

Figure 4.4 – Message screen after login

We have introduced dynamic content for our login operation, which means the application will serve pages based on certain conditions, in this case, the successful validation of the username/password combination. To achieve a successful validation, enter `admin/admin` as the username/password combination, as this exists in the database.

Let's explore the code a bit further to understand how it works:

```
 1    package main
 2
 3    import (
 4        "fmt"
 5        "github.com/gorilla/mux"
 6        "html/template"
 7        "log"
 8        "net/http"
 9        "os"
10        "path/filepath"
11        "time"
12    )
13
14    type staticHandler struct {
15        staticPath string
16        indexPage  string
17    }
```

```go
18
19    func (h staticHandler) ServeHTTP(w
          http.ResponseWriter, r *http.Request) {
20      path, err := filepath.Abs(r.URL.Path)
21      log.Println(r.URL.Path)
22      if err != nil {
23          http.Error(w, err.Error(),
                      http.StatusBadRequest)
24          return
25      }
26
27      path = filepath.Join(h.staticPath, path)
28
29      _, err = os.Stat(path)
30
31      http.FileServer(
            http.Dir(h.staticPath)).ServeHTTP(w, r)
32    }
33
34    func postHandler(w http.ResponseWriter,
                      r *http.Request) {
35      result := "Login "
36      r.ParseForm()
37
38      if validateUser(r.FormValue("username"),
                      r.FormValue("password")) {
39          result = result + "successfull"
40      } else {
41          result = result + "unsuccessful"
42      }
43
44      t, err :=
            template.ParseFiles("static/tmpl/msg.html")
45
46      if err != nil {
47          fmt.Fprintf(w, "error processing")
```

```
48          return
49      }
50
51      tpl := template.Must(t, err)
52
53      tpl.Execute(w, result)
54  }
55
56  func validateUser(username string,
                      password string) bool {
57      return (username == "admin") &&
          (password == "admin")
58  }
59
60  func main() {
61      router := mux.NewRouter()
62
63      router.HandleFunc("/login",
          postHandler).Methods("POST")
64
65      spa := staticHandler{staticPath: "static",
                             indexPage: "index.html"}
66      router.PathPrefix("/").Handler(spa)
67
68      srv := &http.Server{
69          Handler:      router,
70          Addr:         "127.0.0.1:3333",
71          WriteTimeout: 15 * time.Second,
72          ReadTimeout:  15 * time.Second,
73      }
74
75      log.Fatal(srv.ListenAndServe())
76  }
```

The `ServeHTTP` function (line 19) serves the content specified by the directory defined in the `staticHandler` struct (line 65), which points to the `static` directory with the index page showing as `index.html`. The handler configuration is registered using the Gorilla Mux attached to the `/` path prefix (line 66).

The next part is the code that takes care of the registration of the `/login` endpoint (line 63). The `postHandler` function (line 34) extracts and validates the username and password information passed from the request.

The web page contains two input elements, the username and password, which are sent by the browser when the user clicks on the **Login** button. When the handler (line 34) receives the data, it parses it using the `ParseForm()` function (line 36) and then extracts the value passed by referencing the field names `username` and `password` (line 38), which corresponds to the name of the HTML element specified inside the file in `chapter04/dynamic/static/index.html`.

On completing the validation process, the app then uses the Go `html/template` package (line 44) to parse another HTML file (`static/tmpl/msg.html`). The app will parse the HTML file and will insert all the relevant information to be included as part of the HTML page using the `template.Must` function (line 51).

This `msg.html` file contains a `{{.}}` placeholder string that is understood by the `html/template` package (line 18):

```
1      <!DOCTYPE html>
2      <html>
3        <head>
...
18               <p class="text-xs text-gray-50">{{.}}
                 </p>
...
24     </html>
```

In this section, we have learned how to render dynamic content. In the next section, we will look at bundling both our static and dynamic content to allow us to run the application as a single file.

Using Go embed to bundle your content

In this section, we will look at how to package applications into a single binary. Packaging everything the application needs into a single binary makes it easier to deploy the application anywhere in the cloud. We are going to use the `embed` package that is provided by the *Go standard library*. The following link provides further detail on the different functions available inside the embed package: `https://pkg.go.dev/embed`.

> **Note**
> The embed package is only available in Go version 1.16 and upwards.

The following code provides a simple example of using the embed package in three different ways – to embed a specific file, embed the full contents of a folder, and embed a specific file type:

```
1    package main
2
3    import (
4        "embed"
5        "fmt"
6        "github.com/gorilla/mux"
7        "html/template"
8        "io/fs"
9        "log"
10       "net/http"
11       "os"
12       "path/filepath"
13       "strings"
14       "time"
15   )
16
17   var (
18       Version string = strings.TrimSpace(version)
19       //go:embed version/version.txt
20       version string
21
22       //go:embed static/*
23       staticEmbed embed.FS
24
25       //go:embed tmpl/*.html
26       tmplEmbed embed.FS
27   )
28
29   type staticHandler struct {
30       staticPath string
31       indexPage  string
```

```
32    }
33
34    func (h staticHandler) ServeHTTP(w
           http.ResponseWriter, r *http.Request) {
35        path, err := filepath.Abs(r.URL.Path)
36        log.Println(r.URL.Path)
37        if err != nil {
38            http.Error(w, err.Error(),
                        http.StatusBadRequest)
39            return
40        }
41
42        path = filepath.Join(h.staticPath, path)
43
44        _, err = os.Stat(path)
45
46        log.Print("using embed mode")
47        fsys, err := fs.Sub(staticEmbed, "static")
48        if err != nil {
49            panic(err)
50        }
51
52        http.FileServer(http.FS(fsys)).ServeHTTP(w,
                                                    r)
53    }
54
55    //renderFiles renders file and push data (d) into
      // the templates to be rendered
56    func renderFiles(tmpl string, w
        http.ResponseWriter, d interface{}) {
57        t, err := template.ParseFS(tmplEmbed,
            fmt.Sprintf("tmpl/%s.html", tmpl))
58        if err != nil {
59            log.Fatal(err)
60        }
61
```

```
62        if err := t.Execute(w, d); err != nil {
63            log.Fatal(err)
64        }
65    }
66
67    func postHandler(w http.ResponseWriter,
                        r *http.Request) {
68        result := "Login "
69        r.ParseForm()
70
71        if validateUser(r.FormValue("username"),
                        r.FormValue("password")) {
72            result = result + "successfull"
73        } else {
74            result = result + "unsuccessful"
75        }
76
77        renderFiles("msg", w, result)
78    }
79
80    func validateUser(username string,
                        password string) bool {
81        return (username == "admin") &&
                (password == "admin")
82    }
83
84    func main() {
85        log.Println("Server Version :", Version)
86
87        router := mux.NewRouter()
88
89        router.HandleFunc("/login", postHandler)
            .Methods("POST")
90
91        spa := staticHandler{staticPath: "static",
                                indexPage: "index.html"}
```

```
92          router.PathPrefix("/").Handler(spa)
93
94          srv := &http.Server{
95              Handler:        router,
96              Addr:           "127.0.0.1:3333",
97              WriteTimeout: 15 * time.Second,
98              ReadTimeout:    15 * time.Second,
99          }
100
101         log.Fatal(srv.ListenAndServe())
102     }
```

The source code resides inside the chapter4/embed folder. The code uses the //go:embed directive (lines 19, 22, and 25). This tells the compiler that the version string (line 20) will get the content from version/version.txt, which contains the version information that we want to display to the user.

We also declare the //go:embed directive telling the compiler that we want to include everything inside the static/ (line 22) and tmpl/ (line 25) folders. During the compilation process, the compiler detects the preceding directives and automatically includes all the different files into the binary.

The tmpl directory contains the template that will render dynamic content, and since we have embedded it into the binary, we need to use a different way to render it (line 56). The new renderFiles function uses the template.ParseFS function (line 57), which renders the template declared in the tmplEmbed variable.

The renderFiles function is called from the postHandler function (line 77), passing in the template name and other parameters.

Now, this time when building our application, the final executable file contains the different files (HTML, CSS, etc.) in a single file. We can now compile the application, as follows:

go build -o embed

This will generate an executable file – for example, in Linux, it will be called embed and in Windows, it will be called embed.exe. Next, run the application as follows:

./emded

Open your browser and go to http://localhost:3333/. It should look the same as before, except that everything is being retrieved via embed.FS. You now have a fully embedded application that can be deployed as a single binary in the cloud.

Summary

This pretty big chapter served as our first look at interacting with user-provided data and handling web requests. We've seen how we can add RESTful endpoints using the Go standard library and have learned how we can use the utility functions of Gorilla Mux to quickly add more power and functionality to our application. We've also explored the different ways we can handle requests. In one method, we can now utilize Go's `html/template` library to dynamically create content and package it as a directory read from disk. Alternatively, we can use the new Go `embed` directive to give us a single binary that packages up all our assets and makes for simple deployments.

In the next chapter, we will look at adding middleware to help process the request pipeline and introduce security to ensure that content can be accessed securely.

5

Securing the Backend and Middleware

In previous chapters, we learned how to build our database, run our web application as a server, and serve dynamic content. In this chapter, we will discuss security – in particular, we will look at securing the web app. Security is a vast topic so for this chapter, we will just look at the security aspects that are relevant to our application. Another topic that we will look at is middleware and using it as part of our application.

Middleware is software that is introduced into an application to provide generic functionality that is used for incoming and outgoing traffic in our application. Middleware makes it easy to centralize features that are used across different parts of our applications, and this will be discussed more in upcoming sections of this chapter.

In this chapter, we'll be covering the following topics:

- Adding authentication
- Adding middleware
- Adding cookies and sessions with Redis

Upon completing this chapter, you will have learned how to set up a user database and add authentication to the app. We will also learn about middleware and how to add it to an existing app. Lastly, you will learn about cookies, storing information in sessions, and using Redis as persistence storage for these sessions.

Technical requirements

All the source code explained in this chapter can be checked out at `https://github.com/PacktPublishing/Becoming-a-Full-Stack-Go-Developer/tree/main/Chapter05`.

Adding authentication

Building the application requires some consideration in terms of designing the application, and one of the key pieces that needs to be thought of ahead of time is security. There are many facets of security but in this section of our application, we will look at authentication.

> **Note**
> Authentication is the process of validating that a user is who they claim to be.

To add authentication to our app, we will need to store the user information in the database first. The user information will be used to authenticate the user before using the application. The database user table can be found inside the db/schema.sql file:

```
CREATE TABLE gowebapp.users (
User_ID           BIGSERIAL PRIMARY KEY,
User_Name         text NOT NULL,
Password_Hash text NOT NULL,
Name              text NOT NULL,
Config            JSONB DEFAULT '{}'::JSONB NOT NULL,
Created_At        TIMESTAMP WITH TIME ZONE DEFAULT NOW() NOT NULL,
Is_Enabled        BOOLEAN DEFAULT TRUE NOT NULL
```

The following table outlines the data types that are used for the user table:

BIGSERIAL	An auto-incrementing data type that is normally used as a primary key.
TEXT	A variable-length character string.
JSONB	The JSON binary data type is suitable for JSON data. The database provides this data type to make it easier to index, parse, and query JSON data directly.
TIMESTAMP	A date and time data type.
BOOLEAN	A logical data type that contains true or false.

The authentication will be performed by checking the User_Name and Pass_Word_Hash fields. One thing to note – the Pass_Word_Hash field contains an encrypted password, and we will look further into encrypting the password a bit later.

As discussed in *Chapter 1, Building the Database and Model*, we are using sqlc to generate the Go code that will talk to the database. To generate the Go code, execute the following command:

```
make generate
```

The code that will read the user information will be stored under the gen/query.sql_gen.go file as shown here:

```
...
func (q *Queries) GetUserByName(ctx context.Context, userName
string) (GowebappUser, error) {
  row := q.db.QueryRowContext(ctx, getUserByName, userName)
  var i GowebappUser
  err := row.Scan(
      &i.UserID,
      &i.UserName,
      &i.PasswordHash,
      &i.Name,
      &i.Config,
      &i.CreatedAt,
      &i.IsEnabled,
  )
  return i, err
}
...
```

The GetUserByName function queries the database by calling the QueryRowContext() function, passing in the query that we want to use, which is defined as shown here:

```
const getUserByName = `-- name: GetUserByName :one
SELECT user_id, user_name, pass_word_hash, name, config,
created_at, is_enabled
FROM gowebapp.users
WHERE user_name = $1
`
```

The query uses the WHERE clause and expects one parameter, which is the user_name field. This is populated by passing the userName parameter into the QueryRowContext() function.

We will look at how to create a dummy user when we start the application in the next section. A dummy user is a user that is normally used for testing purposes – in our case, we want to create a dummy user to test the authentication process.

Creating our dummy user

Our database is empty so we will need to populate it with a dummy user and in this section, we will look at how to create one. We will add code to create a dummy user when the application starts up. The following function inside `main.go` creates the dummy user, and this user will be used to log in to the application:

```
func createUserDb(ctx context.Context) {
   //has the user been created
   u, _ := dbQuery.GetUserByName(ctx, "user@user")

   if u.UserName == "user@user" {
      log.Println("user@user exist...")
      return
   }
   log.Println("Creating user@user...")
   hashPwd, _ := pkg.HashPassword("password")
   _, err := dbQuery.CreateUsers(ctx,
                                 chapter5.CreateUsersParams{
      UserName:      "user@user",
      PassWordHash: hashPwd,
      Name:          "Dummy user",
   })
   ...
}
```

When the application starts up it will first check whether an existing test user exists and if none exists, it will automatically create one. This is put inside the application to make it easier for us to test the application. The `createUserDb()` function uses the `CreateUsers()` generated sqlc function to create the user.

One of the things you will notice is the password is created by the following code snippet:

```
hashPwd, _ := pkg.HashPassword("password")
```

The password is passed to a `HashPassword` function that will return a hashed version of the clear text password.

The `HashPassword` function uses the Go `crypto` or `bcrypt` standard libraries that provide a function to return a hash of a plain string as shown here:

```
func HashPassword(password string) (string, error) {
```

```
    bytes, err :=
        bcrypt.GenerateFromPassword([]byte(password), 14)
    return string(bytes), err
}
```

The hash generated from the string password will be different whenever the `bcrypt.GenerateFromPassword` function is called. The `GenerateFromPassword()` function uses the standard cryptography library to generate the hash value of the password.

Cryptography is the practice of ensuring text messages are converted into a form that is not easy to read or deconstruct. This provides data security to make it hard to deconstruct what the data is all about. Go provides a standard library that provides cryptography functions, which is available in the `golang.org/x/crypto` package. The `crypto` library provides a number of cryptography functions that you can choose from – it all depends on what you need for your application. In our example, we use `bcrypt`, which is a password-hashing function.

Now that we have added a function to create a dummy user in the database, in the next section, we will look at how to authenticate with the database.

Authenticating a user

User authentication is simple, as the application will use the function generated by sqlc, as shown here:

```
func validateUser(username string, password string) bool {
    ...
    u, _ := dbQuery.GetUserByName(ctx, username)
    ...
    return pkg.CheckPasswordHash(password, u.PassWordHash)
}
```

The `GetUserByName` function is used, with the username passed as a parameter to obtain the user information. Once that has been retrieved successfully, it will check whether the password is correct by calling `CheckPasswordHash`.

The `CheckPasswordHash` function uses the same `crypto` or `bcrypt` package and it calls the `CompareHashAndPassword` function, which will compare the hashed password with the password sent by the client. The function returns `true` if the password matches.

```
func CheckPasswordHash(password, hash string) bool {
    err := bcrypt.CompareHashAndPassword([]byte(hash),
                                         []byte(password))
    return err == nil
}
```

The `validateUser` function will return `true` if the username and password combination exists in the database and is correct.

Start your application and navigate your web browser to `http://127.0.0.1:3333/` and you should see a login prompt. Try logging in with incorrect credentials before entering `user@user / password` – you should now be sent to the successful login screen! Congratulations – you successfully authenticated!

In the next section, we will look at middleware, what it is, and how to add it to our application.

Adding middleware

Middleware is a piece of code that is configured as an HTTP handler. The middleware will pre-process and post-process the request, and it sits between the main Go server and the actual HTTP handlers that have been declared.

Adding middleware as part of our application helps take care of tasks that are outside of the main application features. Middleware can take care of authentication, logging, and rate limiting, among other things. In the next section, we will look at adding a simple logging middleware.

Basic middleware

In this section, we are going to add a simple basic middleware to our application. The basic middleware is shown in the following code snippet:

```
func basicMiddleware(h http.Handler) http.Handler {
    return http.HandlerFunc(func(wr http.ResponseWriter,
                                      req *http.Request) {
        log.Println("Middleware called on", req.URL.Path)
        // do stuff
        h.ServeHTTP(wr, req)
    })
}
```

Gorilla Mux makes it incredibly easy to use our middleware. This is done by exposing a function on the router called `Use()`, which is implemented with a variadic number of parameters that can be used to stack multiple pieces of middleware to be executed in order:

```
func (*mux.Router).Use(mwf ...mux.MiddlewareFunc)
```

The following code snippet shows how we implement the Use() function to register the middleware:

```
func main() {
    ...
    // Use our basicMiddleware
    router.Use(basicMiddleware)
    ...
}
```

mux.MiddwareFunc is simply a type alias for func(http.Handler) http.Handler so that anything that meets that interface can work.

To see our function in action, we simply call router.Use(), pass in our middleware, navigate to our web app, and there we can see that it is called:

```
go build && ./chapter5
2022/01/24 19:51:56 Server Version : 0.0.2
2022/01/24 19:51:56 user@user exists...
2022/01/24 19:52:02 Middleware called on /app
2022/01/24 19:52:02 Middleware called on /css/minified.css
...
```

You may be wondering why you can see it being called multiple times with different paths – the reason is that when requesting our app, it's performing a number of GET requests for the numerous hosted resources. Each of these is passing through our middleware as shown in *Figure 5.1*:

Figure 5.1 – Request passing through middleware

The handlers library – available at https://github.com/gorilla/handlers – contains many other useful middleware methods and we'll be using some of them later, including the handlers. CORS() middleware to allow us to handle **Cross-Origin Resource Sharing (CORS)**. We will look at CORS and using this middleware in more detail in *Chapter 9, Tailwind, Middleware, and CORS*.

In this section, we learned about middleware, the different functionality that it can provide, and how to add it to an app. In the next section, we will look at session handling and using cookies to track user information as they use the application.

Adding cookies and sessions

In this section, we are going to take a look at how we are going to keep track of the users when using our application. We are going to take a look at session management and how it can help our application understand whether a user is allowed to access our application. We are also going to take a look at cookies, which are a session management tool that we are going to use.

The session management discussed in this chapter is part of the Gorilla project, which can be found at `https://github.com/gorilla/sessions`.

Cookies and session handling

In this section, we are going to look at session handling and how to use it to store information relevant to a particular user. The web as we know is stateless in nature, which means that requests are not actually tied to any other previous requests. This makes it hard to know which requests belong to which user. Hence, the need arises to keep track of this and store information about the user.

> **Note**
> A web session is used to facilitate interaction between users and the different services that are used in the sequence of requests and responses. The session is unique to a particular user.

Sessions are stored in memory, with each session belonging to a particular user. Session information will be lost if the application stops running or when the application decides to remove the session information. There are different ways to store session information permanently in storage to be used at a future time.

Figure 5.2 shows the high-level flow of how a session is created and used for each incoming request. New sessions are created when one does not exist and once one is made available, the application can use it to store relevant user information.

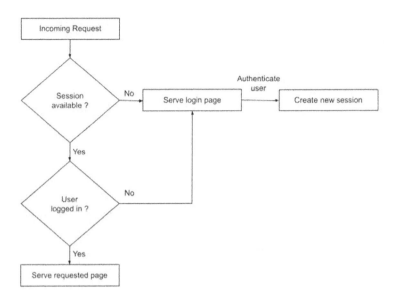

Figure 5.2 – Session check flow

We know that a session is used to store user-specific information – the question is how the application knows which session to use for which user. The answer is a key that is sent back and forth between the application and the browser. This key is called a session key, which is added to the cookie header as shown in *Figure 5.3*.

Figure 5.3 – Cookie containing a session token

As seen in *Figure 5.3*, the cookie with the `session_token` label contains the key that will be sent back to the server to identify the user stored in the session. *Figure 5.3* shows the developer console of the browser. For Firefox, you can open it using the **Tools** > **Web Developer** > **Web Developer Tool** menu, and if you are using Chrome, you can access it using *Ctrl + Shift + J*.

The following snippet shows the `sessionValid` function, which checks whether the incoming request contains a valid `session_token` key. The `store.Get` function will automatically create a new one if an existing session is not available for the current user:

```
//sessionValid check whether the session is a valid session
func sessionValid(w http.ResponseWriter, r *http.Request) bool
{
    session, _ := store.Get(r, "session_token")
    return !session.IsNew
}
```

Once the application finds a session for the user, it will check the authentication status of the user as shown here. The session information is stored as a map, and the map type stores information as key and value, so in our case, we are checking whether the session contains the `authenticated` key:

```
func hasBeenAuthenticated(w http.ResponseWriter, r *http.
Request) bool {
    session, _ := store.Get(r, "session_token")
    a, _ := session.Values["authenticated"]


    . . .
}
```

If there is a failure to obtain the `authenticated` key, the application will automatically redirect the request to display the login page as shown here:

```
//if it does have a valid session make sure it has been
//authenticated
if hasBeenAuthenticated(w, r) {

    . . .
}


//otherwise it will need to be redirected to /login
. . .
http.Redirect(w, r, "/login", 307)
```

We have learned about sessions and how we can use them to check whether a user has been authenticated. We will explore this further.

Storing session information

In the previous section, we learned about sessions and cookie handling. In this section, we will look at how to store session information pertaining to the user. The information stored inside the session is stored in the server memory, which means that this data will be temporarily available as long as the server is still running. Once the server stops running, all the data stored in memory will not available anymore. This is why we will look at persisting the data in a separate storage system in the next section.

In our sample application, we are storing information on whether the user has been authenticated successfully. Users are allowed to access other parts of the application only when they have been successfully authenticated.

Run the sample application and open your browser in private mode (Firefox) or incognito mode (Chrome) and type `http://localhost:3333/dashboard.html` as the address. The application will redirect you to the login page because the session does not exist. The operation to check for the existence of the `authenticated` key is performed inside the `storeAuthenticated` function shown here:

```
func storeAuthenticated(w http.ResponseWriter, r *http.Request,
v bool) {
  session, _ := store.Get(r, "session_token")

  session.Values["authenticated"] = v
  err := session.Save(r, w)

  ...
}
```

The `session.Save` function saves the session into memory after creating the `authenticated` key with a new value pass as part of the function call.

Using Redis for a session

As discussed in the previous section, the sessions are stored in memory. In this section, we will look at storing the session information permanently using Redis. The code samples for this section can be found at `https://github.com/PacktPublishing/Full-Stack-Web-Development-with-Go/tree/main/Chapter05-redis`.

The reason why we want to use Redis is because of its simplicity in terms of data storage, only containing key values. It also can be configured for both in-memory and permanent external storage. For our application, we will need to configure `redis` to store information on the disk to make it permanent. Execute the following `make` command to run `redis`:

```
make redis
```

The following is the full Docker command used to run `redis`:

```
docker run -v $(PWD)/redisdata:/data --name local-redis -p
6379:6379 -d redis --loglevel verbose
```

The command runs `redis` using Docker and specifies the `redisdata` local directory as the location of the permanent file storage for the data. To run the sample application, make sure you also run `postgres` using this command:

```
make teardown_recreate
```

Once both `redis` and `postgres` are up and running, you can now run the sample app and use the web application. The following code snippet shows the `initRedis()` function, which takes care of initializing Redis. The function uses two different packages, which you can find at `https://github.com/redis/go-redis` and `https://github.com/rbcervilla/redisstore`. The `go-redis/redis` package contains the driver and API to communicate with Redis while `rbcervilla/redisstore` contains a simple API to read, write, and delete data from Redis:

```
func initRedis() {
  var err error

  client = redis.NewClient(&redis.Options{
     Addr: "localhost:6379",
  })

  store, err = rstore.NewRedisStore(context.Background(),
                                    client)
  if err != nil {
     log.Fatal("failed to create redis store: ", err)
  }

  store.KeyPrefix("session_token")
}
```

Once the initialization has been completed, the `store` variable will be used to write data to and read it from Redis. Inside the `gorilla` library, the `sessions` package automatically uses the configured `client` object to handle all writing and reading of information to and from `redis`.

A new additional handler is added to allow the user to log out from the application as shown in the handler snippet here:

```
func logoutHandler(w http.ResponseWriter, r *http.Request) {
    if hasBeenAuthenticated(w, r) {
        session, _ := store.Get(r, "session_token")
        session.Options.MaxAge = -1
        err := session.Save(r, w)
        if err != nil {
            log.Println("failed to delete session", err)
    }
    }

    http.Redirect(w, r, "/login", 307)
}
```

The logout operation is done by setting the `Options.MaxAge` field for a session. This indicates to the library that the next time the same `session_token` is passed to the server, it is considered an invalid/expired session and it will redirect to the login page.

Summary

In this chapter, we learned about a few new things that can help our application better. We learned how to add an authentication layer to our application to secure it, which helps protect our application from being accessed anonymously. We also looked at adding middleware to our application and showed how easy it was to add different middleware to our application without changing much code.

Lastly, we looked at session handling and learned how to use it to track user information and a user's journey with our application. Since session handling is not stored permanently, we looked at using the `redis` data store to store the user session information, which allows the application to remember user information anytime the application is restarted.

In the next chapter, we will look at writing code that will process information back and forth between the browser and our application. We will look at building a REST API that will be used to perform different operations on our data.

6

Moving to API-First

In the previous chapters, we learned about building databases, adding monitoring to applications, using middleware, and session handling. In this chapter, we will learn about building an API in our application, and why an API is an important part of writing applications as it forms the interface between the frontend and the backend. Building the API first is important, as it forms the bridge for data exchanges and can be thought of as a contract between the frontend and the backend. Having the proper and correct form of contract is important before building an application.

We will also explore the concepts of REST and JSON to get a better understanding of what they are and how they are used throughout our application.

Upon completion of this chapter, you will know how to design a REST API using Gorilla Mux and also how to process requests to perform operations by converting data to and from JSON. You will also learn how to take care of error handling.

In this chapter, we'll be covering the following topics:

- Structuring API-first applications
- Exposing REST APIs
- Converting data to and from JSON using Go
- Error handling using JSON

Technical requirements

All the source code explained in this chapter can be checked out from `https://github.com/PacktPublishing/Full-Stack-Web-Development-with-Go/tree/main/Chapter06`.

Structuring an application

Go applications are structured inside directories, with each directory containing Go source code that means something for those applications. There are many ways to structure your Go application in different kinds of directories; however, one thing that you have to remember is to always give a directory a name that will be easy for others to understand. As an application grows with time, the chosen directory structure and where code is placed has a big impact on how easily other developers in your team will be able to work with the code base.

Defining packages

Up to this point, we've kept things fairly simple, but we're going to up our game a little and move to a fairly common layout. We won't use the term "standard layout," as there's no such thing in Go, but we'll look at how we're structuring our new project and talk about how we reason them through to best structure our Go application for clarity and understanding, as shown in *Figure 6.1*.

Figure 6.1: Chapter 6 package structure

Let's examine some of these files in a bit more detail to understand these decisions.

generate.go

If you take a look at this file, it can appear confusing at first, but we've used a neat Go feature called go generate that can help:

```
package main
//go:generate echo Generating SQL Schemas
//go:generate sqlc generate
```

At a glance, it looks like a comment because comments in Go start with the // character. However, this one starts with the word go:generate. This is called the go:generate directive; what this means is that when go generate is executed (as shown in the following code block), it will execute the command specified – in our example, it will print the text Generating SQL Schemas and execute the sqlc command-line tool (sqlc generate):

```
$ go generate
Generating SQL Schemas
$
```

This is a useful technique to easily generate your build prerequisites; this can be done as part of your workflow, performed by Makefile, or done by your CI/CD. Makefile is a file containing sets of rules to determine which parts of a program need to be compiled and what command to use to compile the source code. It can be used to compile all kinds of programming language source code.

All we're doing in our generate.go file is simply ensuring that we generate the latest schema files for sqlc. We could add mock generation, more informational messages, or generate archives or any manner of other useful things that might make up our build.

handlers.go

This name comes purely from our experience in using the same pattern of naming files after the functionality defined therein. Our handlers file provides a single place (for now) where our HTTP handlers can be found. Ours contains login, logout, and all kinds of handlers and their request and response types needed to interact with our app. We don't do anything outside of our handlers in this file; all connectivity and addition of middleware are performed as part of main.go to ensure the separation of concerns.

internal/

In the "old days" of Go – back before 1.0 was released – the Go source code featured a directory called pkg, which was for internal-only code and became an idiom for the community, as well as a way to mark subfolders/packages as internal to a particular project.

The pkg folder was eventually removed from the Go project but it left a bit of an unfulfilled need, and to that end, the `internal` directory was created. `internal` is a special directory in so much as it is recognized by the Go tool itself, which allows an author to restrict importing the package unless they share a common ancestor. To demonstrate this, we're storing our API package here as well as `env.go` (used to simplify a way to read environmental variables in the app) and `auth.go` (our specific way to handle authorization) – the `auth.go` or `handlers.go` files in particular are good options to prevent others from importing, while others like the `env` package are more general and can be moved up and out.

migrations, queries, and store

Using `sqlc` and `golang-migrate`, we've given ourselves a leg up in making things easy to organize and increasing our ability to rapidly create our apps. We're just separating things to make life a bit easier, as shown in the `sqlc.yaml` configuration file here:

```
path: store/
schema: migrations/
queries: queries/
```

To see how this works in practice, take a look at the `readme` file provided in the repo.

We have looked at structuring applications by separating different parts of an application into different folders. Grouping source code into different folders allows easier navigation of the application when doing maintenance and development. In the next section, we will explore building an API that will be used to consume data.

Exposing our REST API

Let's understand a few concepts that we are going to use in this section:

- REST – **REST** stands for **Representational State Transfer**. It is a widely accepted set of guidelines for creating web services. REST is independent of the protocol used, but most of the time, it is tied to the HTTP protocol that normal web browsers use. Some of the design principles behind REST include the following:

  - A resource has an identifier – for example, the URI for a particular order might be `https://what-ever-shop.com/orders/1`.

  - Uses JSON as the exchange format – for example, a GET request to `https://what-ever-shop.com/orders/1` might return the following response body:

    ```
    {"orderId":1,"orderValue":0.99,"productId":100,
    "quantity":10}
    ```

- REST APIs built on HTTP are called using standard HTTP verbs to perform operations on resources. The most common operations are GET, POST, PUT, PATCH, and DELETE.

- API – **API** is an acronym for **Application Programming Interface**, a software intermediary that allows two applications to talk to each other. For example, if you are using the Google search engine, you are using an API that it provides.

Combining both the preceding concepts, we come up with a REST API, and the software that we are building is called a RESTful API, which means that the API that we provide can be accessed using REST.

In this section, we will look at exposing our RESTful handlers, a pattern for an API server, and discuss our new `middleware.Main` session and the API package.

We've done some rework to prepare our new API-first project. We've abstracted the API server into its own package in `internal/api`. Its responsibility is to provide a server that accepts a port to bind on and the ability to start the server, stop it, and add routes with optional middleware.

The following is a snippet (from `chapter06/main.go`) of our new main function showing this approach:

```
1    func main() {
2        ...
3        server := api.NewServer(internal.GetAsInt(
                                  "SERVER_PORT", 9002))
4
5        server.MustStart()
6        defer server.Stop()
7
8        defaultMiddleware := []mux.MiddlewareFunc{
9            api.JSONMiddleware,
10           api.CORSMiddleware(internal.GetAsSlice(
                 "CORS_WHITELIST",
11               []string{
12                   "http://localhost:9000",
13                   "http://0.0.0.0:9000",
14               }, ","),
15           ),
16       }
17
18       // Handlers
19       server.AddRoute("/login", handleLogin(db),
```

```
                        http.MethodPost, defaultMiddleware...)
20          server.AddRoute("/logout", handleLogout(),
                        http.MethodGet, defaultMiddleware...)
21
22          // Our session protected middleware
23          protectedMiddleware :=
                append(defaultMiddleware,
                        validCookieMiddleware(db))
24          server.AddRoute("/checkSecret",
                checkSecret(db), http.MethodGet,
                protectedMiddleware...)
25
26          ...
27      }
```

Pay special attention to how we've created our default middleware, which is declared in the `defaultMiddleware` variable (line 8). For our protected routes, we are appending the `protectedMiddleware` variable (line 23) into the existing `defaultMiddleware` variable. Our custom session verification middleware is added to the middleware chain (line 23) to ensure a valid login before allowing access to our other handlers.

We've also pushed two types of middleware into this `api` package, `JSONMiddleware` (line 9) and `CORSMiddleware` (line 10), which takes a slice of strings for a **CORS** allow-list, which we'll look at in more depth in the next section.

Cross-Origin Resource Sharing (CORS)

Anyone working with API-first applications will encounter the concept of CORS. It's a security feature of modern browsers to ensure that web apps on one domain have permission to request APIs on a different origin. The way it does this is by performing what is called a preflight request, which is basically just a normal `OPTIONS` request. This returns information, telling our app that it is allowed to talk to the API endpoint, along with the methods it supports and the origins. Origins contain the same domain sent by the client in the `origin` header, or it could be a wildcard (`*`), which means that all origins are allowed, as explained in *Figure 6.2*.

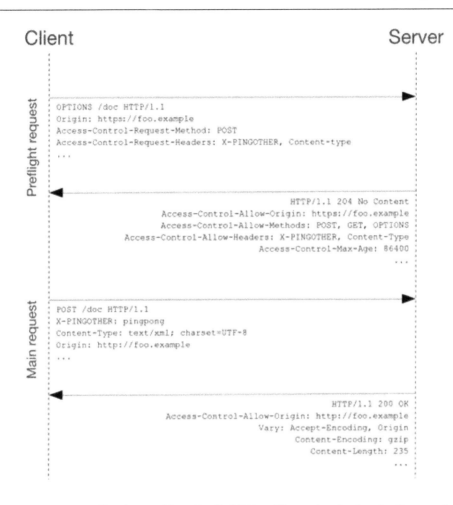

```
OPTIONS /doc HTTP/1.1
Origin: https://foo.example
Access-Control-Request-Method: POST
Access-Control-Request-Headers: X-PINGOTHER, Content-type
...

HTTP/1.1 204 No Content
Access-Control-Allow-Origin: https://foo.example
Access-Control-Allow-Methods: POST, GET, OPTIONS
Access-Control-Allow-Headers: X-PINGOTHER, Content-Type
Access-Control-Max-Age: 86400
...

POST /doc HTTP/1.1
X-PINGOTHER: pingpong
Content-Type: text/xml; charset=UTF-8
Origin: http://foo.example
...

HTTP/1.1 200 OK
Access-Control-Allow-Origin: http://foo.example
Vary: Accept-Encoding, Origin
Content-Encoding: gzip
Content-Length: 235
...
```

Figure 6.2: CORS flow (sourced from Mozilla MDN and licensed under Creative Commons)

Our middleware wraps the Gorilla Mux CORS middleware to make it a little easier for us to provide our CORS whitelisted domains (the domains we're happy to respond to requests on) and all the HTTP methods for those same domains.

JSON middleware

Another piece of middleware that is functionally needed to enforce our requirements for an API-powered application is JSON middleware. **JSON**, short for **Javascript Object Notation**, is an open standard file format that is used to represent data in a key-value pair and arrays.

JSON middleware uses HTTP headers to check what kind of data is being sent in a request. It checks the Content-Type header key, which should contain the application/json value.

If it cannot find the value that it requires, then the middleware will check the value of the Accept header to see whether it can find the application/json value. Once the check is done and it cannot find the value that it is looking for, it replies that it's not a suitable content type for us to work with. We also add that header to our ResponseWriter so that we can ensure we're telling the consumer we only support JSON and send that back to them.

The following code snippet shows the JSON middleware:

```
1   func JSONMiddleware(next http.Handler)
       http.Handler {
2       return http.HandlerFunc(func(wr
           http.ResponseWriter, req *http.Request) {
3           contentType :=
               req.Header.Get("Content-Type")
4
5           if strings.TrimSpace(contentType) == "" {
6               var parseError error
7               contentType, _, parseError =
                   mime.ParseMediaType(contentType)
8               if parseError != nil {
9                   JSONError(wr,
                       http.StatusBadRequest,
                       "Bad or no content-type header
                       found")
10                  return
11              }
12          }
13
14          if contentType != "application/json" {
15              JSONError(wr,
                   http.StatusUnsupportedMediaType,
                   "Content-Type not
                   application/json")
16              return
17          }
18          // Tell the client we're talking JSON as
            // well.
19          wr.Header().Add("Content-Type",
```

```
                                    "application/json")
20              next.ServeHTTP(wr, req)
21          })
22      }
```

Line 14 checks whether the Content-Type header contains an application/json value; otherwise, it will return an error as part of the response (line 15).

Now that we understand the concept of middleware, we'll develop some middleware to make handling our sessions easier.

Session middleware

This session middleware does not fit into our api package as it's closely tied to our session-handling functionality, as shown in the following code snippet:

```
1      session, err := cookieStore.Get(req,
                                    "session-name")
2      if err != nil {
3          api.JSONError(wr,
                        http.StatusInternalServerError,
                        "Session Error")
4          return
5      }
6
7      userID, userIDOK :=
          session.Values["userID"].(int64)
8      isAuthd, isAuthdOK :=
          session.Values["userAuthenticated"].(bool)
9      if !userIDOK || !isAuthdOK {
10          api.JSONError(wr,
              http.StatusInternalServerError,
              "Session Error")
11          return
12      }
13
14      if !isAuthd || userID < 1 {
15          api.JSONError(wr, http.StatusForbidden,
                        "Bad Credentials")
```

```
16        return
17    }
18    ...
19    ctx := context.WithValue(req.Context(),
                            SessionKey, UserSession{
20        UserID: user.UserID,
21    })
22    h.ServeHTTP(wr, req.WithContext(ctx))
23
```

What the preceding middleware does is attempt to retrieve our session from `cookiestore` (line 1), which we covered in the previous chapter. From the returned session map, we perform an assertion on two values (line 7) that assigns `userID` the `int64` value and the Boolean `userIDOK`.

Finally, if everything checks out, including a check of the database for the user, we use `context.WithValue()` (line 19) to provide a new context with our `sessionKey`, which is a unique type to our package.

We then provide a simple function called `userFromSession` that our handlers can call to check the validity of the key type and the incoming session data.

In this section, we learned about middleware and looked at adding different types of middleware to an application. Also, we looked at CORS and how it works when developing web applications. In the next section, we will look in more detail at JSON and use models to represent JSON for requests and responses.

Converting to and from JSON

In this section, we will look at getting and sending data from and to JSON. We will also look at creating a structure to handle data and how the JSON conversion is done.

When dealing with JSON in Golang via the standard library, we've got two primary options – `json.Marshal/Unmarshal` and `json.NewEncoder(io.Writer)/NewDecoder(io.Reader)`. In this chapter, we will look at using the `Encoder/Decoder` methods. The reason for using these methods is that we can chain a function to the encoder/decoder that's returned and call the `.Encode` and `.Decode` functions with ease. Another benefit of this approach is that it uses the streaming interface (namely `io.Reader` and `io.Writer`, used to represent an entity from/to which you can read or write a stream of bytes – the `Reader` and `Writer` interfaces are accepted as input and output by many utilities and functions in the standard library), so we have other choices than `Marshal`, which works with preallocated bytes, meaning we're more efficient with our allocations and also faster.

Defining request model

Data that flows through our application will be wrapped inside a struct. A struct is a structure that is defined to hold data. This makes it easier to transport data across different parts of the application; it does not make sense, if you have to transport 10 different pieces of data to different parts of the application, to do this by calling a function with 10 parameters, but if it is inside a struct, the function will only have to accept one parameter of that type. For simplicity, structs that hold data are also called models, as the field defined inside the struct is modeled on the data that it represents.

Let's take a look at the model that we defined to wrap the login data (username and password) in the following code:

```go
func handleLogin(db *sql.DB) http.HandlerFunc {
  return http.HandlerFunc(func(wr http.ResponseWriter, req
    *http.Request) {

    type loginRequest struct {
      Username string `json:"username"`
      Password string `json:"password"`
    }
    ...
}
```

As seen in the preceding code, the `loginRequest` model is declared with a `json:"username"` definition. This tells the standard library JSON converter the following:

- `username` – the key name used when converted to a JSON string

- `omitempty` – if the value is empty, the key will not be included in the JSON string

More information can be found at `https://pkg.go.dev/encoding/json#Marshal`, where you can see the different configurations that a model can have to convert from/to JSON.

Now that we have defined the model inside the function, we want to use it. The `handleLogin` function uses the Decode function that exists inside the `json` standard library to decode the data, as shown in the following snippet:

```go
payload := loginRequest{}
if err := json.NewDecoder(req.Body).Decode(&payload); err !=
nil {
  ...
}
```

Once successfully converted, the code can use the `payload` variable to access the values that were passed as part of the HTTP request.

Let's take a look at another model that the code defines to store exercise set information that is passed by the user. The way to convert the data into `newSetRequest` is the same as `loginRequest` using the Decode function:

```
1    func handleAddSet(db *sql.DB) http.HandlerFunc {
2     return http.HandlerFunc(func(wr
        http.ResponseWriter,
        req *http.Request) {
3
4       ...
5
6       type newSetRequest struct {
7         ExerciseName string
          `json:"exercise_name,omitempty"`
8         Weight    int `json:"weight,omitempty"`
9       }
10
11      payload := newSetRequest{}
12      if err := json.NewDecoder(req.Body)
          .Decode(&payload); err != nil {
13        ...
14        return
15      }
16
17      ...
18    })
19   }
20
```

The function declares a new struct (line 6) called `newSetRequest`, and this will be populated by calling the `json.NewDecoder()` function (line 12), which will be populated into the `payload` (line 11) variable.

In this section, we looked at using a model to host the information that is passed by the user. In the next section, we will look at sending responses back using the model.

Defining a response model

In this section, we will look at how to use a model to host information that will be sent back as a response to the user. In *Chapter 1, Building the Database and Model*, we learned about sqlc tools that generate the different database models that will be used by our application. We will use the same database model defined by sqlc, converted to a JSON string as a response back to the user. The json package library is smart enough to convert models into JSON strings.

Let's look at the response sent back when a user creates a new workout – in this case, the handleAddSet function, as shown here:

```
func handleAddSet(db *sql.DB) http.HandlerFunc {
  return http.HandlerFunc(func(wr http.ResponseWriter,
                               req *http.Request) {
    ...

    set, err :=
      querier.CreateDefaultSetForExercise(req.Context(),
        store.CreateDefaultSetForExerciseParams{
        WorkoutID:   int64(workoutID),
        ExerciseName: payload.ExerciseName,
        Weight:    int32(payload.Weight),
      })
    ...
    json.NewEncoder(wr).Encode(&set)
  })
}
```

As you can see, the function calls the CreateDefaultSetForExercise function and uses the set variable as a response back to the user by using the Encode function. The returned set variable is of type GowebappSet, which is defined as follows:

```
type GowebappSet struct {
  SetID     int64 `json:"set_id"`
  WorkoutID  int64 `json:"workout_id"`
  ExerciseName string `json:"exercise_name"`
  Weight     int32 `json:"weight"`
  Set1      int64 `json:"set1"`
  Set2      int64 `json:"set2"`
```

```
    Set3     int64 `json:"set3"`
  }
```

When the model is converted using `Encode` and sent back as a response, this is how it will look:

```
{
  "set_id": 1,
  "workout_id": 1,
  "exercise_name": "Barbell",
  "weight": 700,
  "set1": 0,
  "set2": 0,
  "set3": 0
}
```

In this section, we looked at a model generated by sqlc that is not only used to host read/write data to and from a database but also used to send responses back to the user as a JSON string. In the next section, we will look at another important feature that we need to add to the application, error handling, which will be reported using JSON.

Reporting errors with JSON

There are many ways to handle errors when writing web applications. In our sample application, we handle errors to inform users of what's happening with their request. When reporting errors to users about their request, remember not to expose too much information about what's happening to the system. The following are some examples of error messages reported to users that contain such information:

- There is a connection error to the database
- The username and password are not valid for connecting to the database
- Username validation failed
- The password cannot be converted to plain text

The preceding JSON error use cases are normally used in scenarios where more information needs to be provided to the frontend to inform users. Simpler error messages containing error codes can also be used.

Using JSONError

Standardizing error messages is as important as writing proper code to ensure application maintainability. At the same time, it makes it easier for others to read and understand your code when troubleshooting.

In our sample application, we will use JSON to wrap error messages that are reported to the user. This ensures consistency in the format and content of the error. The following code snippet can be found inside the `internal/api/wrappers.go` file:

```
1    func JSONError(wr http.ResponseWriter,
       errorCode int, errorMessages ...string) {
2    wr.WriteHeader(errorCode)
3    if len(errorMessages) > 1 {
4      json.NewEncoder(wr).Encode(struct {
5        Status string  `json:"status,omitempty"`
6        Errors []string `json:"errors,omitempty"`
7      }{
8        Status: fmt.Sprintf("%d / %s", errorCode,
           http.StatusText(errorCode)),
9        Errors: errorMessages,
10     })
11     return
12   }
13
14   json.NewEncoder(wr).Encode(struct {
15     Status string  `json:"status,omitempty"`
16     Error string  `json:"error,omitempty"`
17   }{
18     Status: fmt.Sprintf("%d / %s", errorCode,
           http.StatusText(errorCode)),
19     Error: errorMessages[0],
20   })
21 }
```

The JSONError function will use the passed errorCode parameter and errorMessages (line 1) as part of the JSON reported to the user – for example, let's say we call the /login endpoint with the wrong credentials using the following cURL command:

```
curl http://localhost:9002/login -H 'Content-Type: application/
json' -X POST -d '{"username" : "user@user", "password" :
"wrongpassword"}
```

You will get the following JSON error message:

```
{"status":"403 / Forbidden","error":"Bad Credentials"}
```

The error is constructed by using the struct that is defined when encoding the JSON string (line 14).

Using JSONMessage

The sample application uses JSON not only for reporting error messages but also for reporting successful messages. Let's take a look at the output of a successful message. Log in using the following cURL command:

```
curl http://localhost:9002/login -v -H 'Content-Type:
application/json' -X POST -d '{"username" : "user@user",
"password" : "password"}'
```

You will get output that looks like this:

```
*   Trying ::1:9002...
* TCP_NODELAY set
* Connected to localhost (::1) port 9002 (#0)
> POST /login HTTP/1.1
> Host: localhost:9002
...
< Set-Cookie: session-name=MTY0NTM0OTI1OXxEdi1CQkFFQ1
80SUFBUkFCRUFBQVJQLUNBQU1HYzNSeWFFXNW5EQk1BRVhWelpYSkJ
kWFJvW1c1MGFXTmhkR1ZrQkdz.KdmIyd0NBZ0FCQm55OMGNtbHVad3dJ-
QUFaMWMyVn1TVVFGYVc1ME5qUUVBZ0FDQ1HHy75qzLVPoMZ3BbNY17qBWd_
puOh16jpgY-d29ULUV; Path=/; Expires=Sun, 20 Feb 2022 09:42:39
GMT; Max-Age=900; HttpOnly
...
* Connection #0 to host localhost left intact
```

Using the `session-name` token, use the following cURL command to create a workout:

```
curl http://localhost:9002/workout -H 'Content-Type:
application/json' -X POST --cookie 'session-name=
MTY0NTM0OTI1OXxEdi1CQkFFQ180SUFBUkFCRUFBQVJQLUNBQU1HYzNSeWFFXNW
5EQk1BRVhWelpYSkJkWFJvW1c1MGFXTmhkR1ZrQkdz.KdmIyd0NBZ0FCQm5yOM
GNtbHVad3dJQUFaMWMyVn1TVVFGYVc1ME5qUUVBZ0FDQ1HHy75qzLVPoMZ3BbNY
17qBWd_puOh16jpgY-d29ULUV'
```

On successfully creating the workout, you will see a JSON message that looks like the following:

```
{"workout_id":3,"user_id":1,"start_date":"2022-02-
20T09:29:25.406523Z"}
```

Summary

In this chapter, we've looked at creating and leveraging our own middleware for session handling as well as enforcing JSON usage on our API. We've also reworked our project to use a common package layout to help separate our concerns and set ourselves up for future work and iteration.

Also in this chapter, we've introduced a number of helper functions, including two for creating and reporting errors and messages to the user via JSON and an API package to abstract our server handling, making it easy to understand and preparing us to accommodate CORS.

In the next chapter, we will discuss writing frontends in more detail and learn how to write frontend applications using a frontend framework.

Part 3: Single-Page Apps with Vue and Go

In *Part 3*, we introduce frontend frameworks before diving into how we can combine Vue with Go and explore different frontend technologies to power our sample applications. We will look at implementing **Cross-Origin Resource Sharing (CORS)** and using JWT for sessions in our application to simplify and secure our app from bad actors!

This part includes the following chapters:

- *Chapter 7, Frontend Frameworks*
- *Chapter 8, Frontend Libraries*
- *Chapter 9, Tailwind, Middleware, and CORS*
- *Chapter 10, Session Management*

7

Frontend Frameworks

In this chapter, we will take a high-level look at the current JavaScript frameworks available to modern web developers. We will compare some of the popular ones, Svelte, React, and Vue, before creating a simple app in Vue and ending by adding navigation using the popular Vue Router. This will lay the foundations needed to later talk to our API server from *Chapter 6, Moving to API-First*.

 Upon completion of this chapter, we will have covered the following:

- Understanding the difference between server-side rendering and single-page applications
- Looking at different frontend frameworks
- Creating applications using the Vue framework
- Understanding routing inside the Vue framework

This chapter paves the way to the land of the frontend. We will learn about the different parts of frontend development in this and the next chapters.

Technical requirements

All the source code used in this chapter can be checked out from `https://github.com/ PacktPublishing/Full-Stack-Web-Development-with-Go/tree/main/Chapter07`.

Make sure you have all the necessary tools installed on your local machine by following the instructions from the Node.js documentation:`https://docs.npmjs.com/downloading-and- installing-node-js-and-npm`.

Server-side rendering versus single-page apps

In *Chapter 4, Serving and Embedding HTML Content*, we created our app as a server-side rendered app. What this means is that all of the content and assets, including the HTML, are generated on the backend and sent on each page request. There's nothing wrong with this; our publisher, Packt, uses **server-side rendering** (**SSR**) for its own site at `https://www.packtpub.com/`. SSR as a technique is used by technologies such as WordPress and many other sites that host content that changes less frequently and may have less interactivity.

The alternative to SSR we're going to use for our app is **client-side rendering** (**CSR**). CSR works by having the client fetch the app as a *bundle* of JavaScript and other assets, executing the JavaScript and the app dynamically, and binding to an element that takes over the page rendering. The app creates and renders each route dynamically in the browser. This is all done without requiring any reloading of the bundle or content.

By moving to client-side rendering, it improves the app's interactivity and responsiveness by allowing it to manipulate the document model, fetch additional content and data via the API, and generally perform closer to what a user might expect from a desktop app without constant page reloads.

When we talk about reactivity, we're describing the situation in which changes in the application state are automatically reflected in the **document object model** (**DOM**). This is a key attribute of all of the frameworks we'll be exploring in this chapter, including React, Vue, and Svelte.

Introducing React, Vue, and more

If there's one thing that the JavaScript community enjoys doing, it's creating new frameworks!

We're going to explore and contrast a few of the most popular ones and look at the common parts they all share and the main points of difference.

React

React is one of the most popular JavaScript libraries available. It was created, and is still maintained, by Meta (formerly Facebook) and was inspired heavily by a predecessor used internally within Facebook for creating PHP components.

React uses the **JavaScript Syntax eXtension** (**JSX**) as a syntax, which looks like a combination of HTML and Java Script. Although you can use React without compilation, most React developers are used to the process common to most modern frameworks, which is to combine and build the source files, the `.jsx` and `.vue` files, and build them into a final bundle that can be deployed as a static file. We'll look at this in a later chapter.

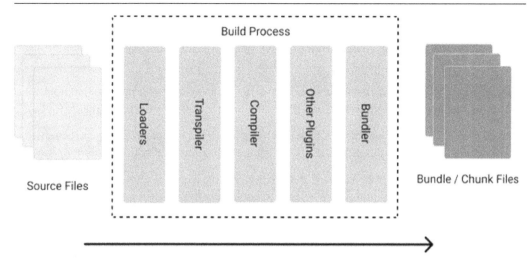

Figure 7.1: Modern JavaScript SPA build process

React is a very popular option for creating apps and one of its strengths is the fact that there are a number of different options to choose from when building your app, such as Redux, Flux, BrowserRouter, or React Router. This flexibility is great but can cause conflict and strong opinions on the "one true way." The React team avoids this issue by continually calling out that React is a library, not a framework, so choosing the components of your app is down to the individual.

React is similar to other frameworks in that it has a full life cycle model that can be "hooked" at runtime to override defaults (for example, `render` and `shouldComponentUpdate`).

Svelte

Svelte straddles an interesting middle ground and is included as an alternative to the two heavy hitters of React and Vue. Svelte takes the approach of pushing more into the compilation step, removing the need for techniques such as diffing the virtual DOM to transpile the code into vanilla JavaScript. This approach means less work is done by the browser but it still has a similar build process to both React and Vue for building bundles. Svelte provides its own preferred router, called SvelteKit, but alternatives exist and Svelte can represent a nice, lightweight alternative to the others. Svelte is quite a new project when looking at the more established players and doesn't have as many resources behind it, but it is still viable for smaller projects.

Vue

The final framework we're going to introduce is Vue, which is what we use as our preferred framework for building our frontend application.

The big appeal initially for me was the fact that the previous version of Vue (version 2) could be loaded and run directly via a **content domain network** (**CDN**), which made experimentation and prototyping incredibly easy back in 2016 when it was first released.

Vue offers a very familiar syntax that makes it easy to learn – it separates out the presentation from the logic and the styling, it's very lightweight, and it uses the concept of **single-file components** (**SFCs**).

The concept of SFC makes it incredibly easy to build simple, scoped components that can be reused from project to project without the addition of learning the "not-quite-JavaScript" JSX used by React.

The following code is a simple component that displays a greeting using the Options API. When Vue was first released, it used the Options API by default, but in later iterations, it has moved to include a newer Composition API, which we'll explore later:

```
<template>
  <div>
    <Thing @click="greetLog" />
    <p class="greeting">{{ greeting }}</p>
  </div>
</template>

<script>
import Thing from '@/components/thing.vue';

export default {
  name: 'Greeter',
  components: ['Thing'],
  props:{},
  mounted(){},
  methods: {
    greetLog() {  console.log('Greeter') };
  },
  data() {
    return {
      greeting: 'Hello World!'
    }
  }
}
</script>
```

```
<style scoped>
.greeting {
  color: red;
  font-weight: bold;
}
</style>
```

Example of a SFC Greeter.vue

As you can see in the preceding code block, the approach of Vue's SFC design has three parts: the HTML, the JavaScript, and the style (usually CSS, often "scoped"). This means you can combine the HTML-esque style of the `<template>` with small Vue-specific additions, such as `@click="functionName"`, to easily create our components. The `@click` annotation featured here, which looks close to HTML, is the syntax used by Vue to extend and bind HTML events to our objects – in this case, replacing the native `onClick` attribute.

The `<script>` contained instance includes a name; props, used to provide properties to the component from parents; `mounted()`, a function called when the component is first added to the DOM; components, that is, the components being imported for use by the component; assorted other methods; and finally, the `data()` object, which can hold our components' state.

The final part of the SFC is the `<style>` part – we can specify non-CSS languages here. For example, we could use `lang="scss"` if we wanted to use SCSS rather than CSS. We can also add the `scoped` keyword, which means that Vue will use name mangling to ensure that our CSS styles are scoped only to this component instance.

A final benefit of using Vue is the opinionated approach taken to build tools (preferring to create Vite, which leverages the incredibly fast esbuild to reduce bundle build times to milliseconds compared to the slower React), component layout, and routers (Vue Router), which we'll explore in later chapters. The opinionated nature of Vue works nicely with the opinionated nature of Golang itself, which helps remove a lot of debate on which approach and components to choose to build your app, ensuring that when you bring in more team members and hand over your successful full stack app, you can be safe in the knowledge that another Vue developer wouldn't argue with you on how you did it, nor on the technology chosen – mainly as they would've chosen the same!

So far in this section, we have looked at what the Vue framework is all about. In the next section, we will learn by creating some simple apps using the Vue framework.

Creating a Vue app

In the previous section, we discussed different frontend frameworks, so for this section, we are going to try to use Vue to build our frontend. In this section, we will look at writing our UI in Vue and discuss how we migrate the login page to Vue. This section will not teach you how to use Vue but rather will look at the way we use Vue to write the frontend components for our sample application.

Application and components

When writing software using Vue, the application will start up by creating an application instance. This instance is the main object in our Vue-based application. Once we have an instance, then we can start using components. Components are reusable UI pieces that contain three parts – a template (which is like HTML), styles, and JavaScript. Normally, when designing a frontend, we think about HTML elements – div, href, and so on – but now we need to think about components that contain all the different parts. *Figure 7.2* shows an example of the login page that we rewrite using Vue.

Figure 7.2: Vue-based login

The concept of an application inside Vue can be thought of as a self-isolated container containing different components that can share data. Any web page can contain a number of applications displaying different kinds of data, and even if they are isolated, they can also share data if and when required.

Login page using Vue

In this section, we will look at how we use the login page as is without converting it into a component and use it as a Vue application rendered by the browser. We need to install the dependencies first by running the following command:

```
npm install
```

This will install all the different dependencies, including the `http-server` module, which we will be using to serve the login page. Start the server by running the following command, making sure you are inside the `chapter7/login` directory:

```
npm run start
```

You will see the output shown in *Figure 7.3*:

```
> test@1.0.0 start
> http-server ./ -p 3000

Starting up http-server, serving ./
Available on:
    http://127.0.0.1:3000
    http://192.168.1.3:3000
    http://172.17.0.1:3000
Hit CTRL-C to stop the server
⊓
```

Figure 7.3: Serving using http-server

Open your browser and type `http://127.0.0.1:3000/login.html` into the address bar, and you will see the login page.

Let's dig through the code and see how it works together. The following snippet inside `login.html` shows the application initialization code:

```
<script type="module">
    import {createApp} from 'vue'

    const app = createApp({
        data() {
            return {
                loginText: 'Login to your account',
                ...
            }
        },
        methods: {
            handleSubmit: function () {
```

```
            . . .
         }
      }
   }).mount('#app')

</script>
```

The code imports `createApp` from the Vue library and uses it to create an application that contains `data()` and `methods` used inside the page. The `data()` block declares the variables that will be used inside the page while `methods` contains functions used. The application is mounted into the element with the ID "app" app, in this case, the `<div>` with `id=app`.

The following code snippet shows the part of the page that uses the data:

```
<body class="bg-gray-900">
      . . .
            <p class="text-xs text-gray-50">{{ loginText
               }}</p>
      . . .
            <p class="text-xs text-gray-50">
               {{ emailText }}</p>
      . . .
            <p class="text-xs font-bold text-white">
               {{ passwordText }}</p>
            . . .
</body>
```

The variable inside the curly brackets ({{ }}) will be populated with the data defined previously when we initialize the application.

The following code snippet shows the part of the page that uses the `handleSubmit` function:

```
<body class="bg-gray-900">
            . . .
               <button @click="handleSubmit"
                     class="px-4 pt-2 pb-2.5 w-full
                           rounded-lg bg-red-500
                           hover:bg-red-600">
```

```
                          . . .
</body>
```

@click on the button element will trigger the function that was defined when creating the Vue application object, which will write to the console log the data in the username field.

Using Vite

Referring back to *Figure 7.1*, one of the parts of the build process is that of the bundler. In this section, we will look at Vite, which is a bundler for Vue. What is a bundler? It is a build tool that combines all your different assets (HTML, CSS, and so on) into one file, making it easy for distribution.

In the previous section, we linked to a CDN-hosted version of the Vue runtime. In this section, we'll be using Vite to build our application and generate our bundled code.

Vite – French for "quick" – was built by the same team behind Vue itself and was designed to provide a faster development experience with extremely fast hot reload and combine it with a powerful build stage that transpiles, minifies, and bundles your code into optimized static assets ready for deployment. Refer back to *Figure 7.1* to see all the stages used to build SPAs.

In this section, we will look at writing our login page as a component and using it as a Vue application rendered by the browser. The code can be seen inside the chapter7/npmvue folder.

Open your terminal and run the following commands:

```
npm install
npm run dev
```

Once the server is up and running, you will get the output shown in *Figure 7.4*.

```
vite v2.8.6 dev server running at:

> Local:   http://localhost:3000/
> Network: use `--host` to expose

ready in 202ms.
```

Figure 7.4: Vite server output

Open the browser and access the login page by entering http://localhost:3000 into the address bar. Let's investigate further and look at how the code is structured. We will start by looking at the index.html page, as shown in the following snippet:

```
<!DOCTYPE html>
<html lang="en">
```

```
<head>
  ...
</head>
<body>
  <div id="app"></div>
  <script type="module" src="/src/main.js"></script>
</body>
</html>
```

The preceding `index.html` references the `main.js` script, which is how we inject the Vue initialization code.

The `<div..>` declaration is where the application will be mounted when rendered in the browser, and the page also includes a script found in `src/main.js`.

`main.js` contains the Vue application initialization code, as shown:

```
import { createApp } from 'vue'
import App from './App.vue'


createApp(App).mount('#app')
```

`createApp` will create an application using the `App` object imported from `App.vue`, which will be the starting component for our application. Vue-related code is normally stored inside a file with the `.vue` extension. The `App.vue` file acts as an app container that hosts the components that it will use. In this case, it will use the `Login` component, as shown in the following snippet:

```
<script setup>
import Login from './components/Login.vue'
</script>


<template>
    <Login />
</template>
```

The `<script setup>` tag is known as the Composition API, which is a set of APIs that allows Vue components to be imported. In our case, we are importing the components from the `Login.vue` file.

The code imports the `Login.Vue` file as a component and uses it inside the `<template>` block. Looking at the `Login.vue` file, you will see that it contains the HTML elements to create the login page.

The `Login.vue` snippet can be seen in the following code block:

```
<script>
export default {
  data() {
    return {
      loginText: 'Login to your account',
      ...
    }
  },
  methods: {
    handleSubmit: function () {
      ...
    }
  }
}
</script>

<style>
@import "../assets/minified.css";
</style>

<template>
    ...
        <button @click="handleSubmit"
              class="px-4 pt-2 pb-2.5 w-full rounded-lg
                    bg-red-500 hover:bg-red-600">
    ...
</template>
```

The class used for the button in the preceding example is declared inside a `minified.css` file inside the `assets` folder.

We have learned how to create apps using the Vue framework and wired all the different components together. We also looked at how to use the Vite tool to write a Vue-based application. In the next section, we will look at routing requests to different Vue components.

Using Vue Router to move around

In this section, we will look at Vue Router and learn how to use it. Vue Router helps in structuring the frontend code when designing a **single-page application (SPA)**. An SPA is a web application that is presented to the user as a single HTML page, which makes it more responsive as the content inside the HTML page is updated without refreshing the page. The SPA requires the use of a router that will route to the different endpoints when updating data from the backend.

Using a router allows easier mapping between the URL path and components simulating page navigation. There are two types of routes that can be configured using Vue Router – dynamic and static routes. Dynamic routes are used when the URL path is dynamic based on some kind of data. For example, in /users/:id, id in the path will be populated with a value, which will be something such as /users/johnny or users/acme. Static routes are routes that do not contain any dynamic data, for example, /users or /orders.

In this section, we will look at static routes. The examples for this section can be found in the chapter7/router folder. Run the following command from the router folder to run the sample application:

```
npm install
npm run server
```

The command will run a server listening on port 8080. Open your browser and enter http://localhost:8080 in the address bar. You will see the output shown in *Figure 7.5*:

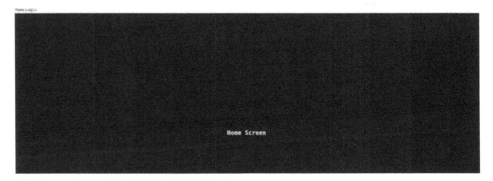

Figure 7.5: Router sample application

The App.vue file contains the Vue Router information, which can be seen as follows:

```
<template>
 <div id="routerdiv">
   <table>
      ...
        <router-link :to="{ name: 'Home'}">Home
```

```
            </router-link>
        ...
            <router-link :to="{ name: 'Login'}">Login
            </router-link>
        ...
    </table>
    <router-view></router-view>
  </div>
</template>
```

The preceding `router-link` route is defined inside `router/index.js`, as shown:

```
const routes = [
    {
        path: '/',
        name: 'Home',
        component: Home
    },
    {
        path: '/login',
        name: 'Login',
        component: Login
    },
];
```

The `<router-link/>` tag defines the router configuration that the application has, and in our case, this is pointing to the Home and Login components declared inside the `index.js` file under the router folder, as shown:

```
import Vue from 'vue';
import { createRouter, createWebHashHistory } from 'vue-router'

import Home from '../views/Home.vue';
import Login from "../views/Login.vue";

Vue.use(VueRouter);

const routes = [
```

```
    {
        path: '/',
        name: 'Home',
        component: Home
    },
    {
        path: '/login',
        name: 'Login',
        component: Login
    },
];

const router = createRouter({
    history: createWebHashHistory(),
    base: process.env.BASE_URL,
    routes
})


export default router
```

Each of the defined routes is mapped to its respective components, which are the Home and Login components, which can be found inside the `views` folder.

Routing the login page

We know that the `/login` path is mapped to the Login component, which is the same component that we looked at in the previous section, *Login page using Vue*. The difference in the router example is in the way the script is defined, as shown:

```
<template>
  ...
</template>
<script type="module">
export default {
  data() {
    return {
```

```
      loginText: 'Login to your account',
      emailText: 'Email Address',
      passwordText: 'Password',
      username: 'enter username',
      password: 'enter password',
    };
  },
  methods: {
    handleSubmit: function () {
      console.log(this.$data.username)
    }
  }
};
</script>
```

Unlike in the previous section, the Vue initialization code has been moved into main.js, as shown:

```
...
const myApp = createApp(App)
myApp.use(router)
myApp.mount('#app')
```

In this section, we looked at how to restructure the application to work as a SPA by using Vue Router.

Summary

In this chapter, we learned about Vue and how to structure our frontend to make it easy to transition into components and applications. We looked at the different frontend frameworks and discussed what each of them provides.

We looked at how components and applications work together when writing a Vue-based web page. We tested what we learned by migrating the login page that we created as a simple HTML page to a Vue-based application. Lastly, we learned about Vue Router and how to use it to make it easier to route to different parts of a SPA.

Taking on board all this learning, in the next chapter, we will look at writing our application as a Vue-based application that will communicate with the REST API that we have built.

8

Frontend Libraries

In the previous chapter, we looked at different frameworks for building frontend applications. In this chapter, we will look at the different frontend libraries that are useful for building web applications. Frontend libraries are predefined functions and classes that can help speed up the development time when building frontend applications by providing functionality we'd otherwise have to build and develop ourselves. In this chapter, we will be looking at the following libraries:

- `Vuetify`
- `Buefy`
- `Vuelidate`
- `Cleave.js`

Having completed this chapter, you will have explored the following:

- Validating data with `Vuelidate`
- Better input handling with `Cleave.js`
- Working with different UI components using `Vuetify`

Technical requirements

All the source code explained in this chapter can be checked out at `https://github.com/PacktPublishing/Full-Stack-Web-Development-with-Go/tree/main/Chapter08`.

Make sure you have all the necessary tools installed on your local machine by following the instructions from the `node.js` documentation available here: `https://docs.npmjs.com/downloading-and-installing-node-js-and-npm`.

In this chapter, there will be sample code that is shared using `codesandbox.io` and `jsfiddle.net`, which will make it easier for you to experiment with.

Let's begin our journey by looking into Vuetify in the next section.

Understanding Vuetify

In *Chapter 7*, *Frontend Frameworks*, we learned about the Vue framework, which is a rich frontend framework that allows frontend code to be extended and maintained easily. Vuetify (`https://vuetifyjs.com`) provides a lot of user interface components out of the box, ready to be used by applications. The framework also allows developers to style the user interfaces to their needs.

In this section, we will learn about Vuetify, which is a Material-based design framework that is built on top of Vue. Material is the Design Language made popular by Google across their web apps and Android applications – you can find out more at `https://m3.material.io/` –and is a very popular choice.

Setting up Vuetify

We are going to take a look at the example code inside the `chapter08/vuetify/components` directory. The example code demonstrates how to use the `Vuetify` framework. Before running the sample code, make sure you run the following command from inside the `chapter08/vuetify/components` directory to install all the necessary components:

```
npm install
```

Once the installation is complete, run the sample code using the following command:

```
npx vue-cli-service serve
```

Once the server is up and running, you will get an output as shown in *Figure 8.1*:

```
DONE  Compiled successfully in 7784ms


App running at:
- Local:   http://localhost:8080/
- Network: http://192.168.1.5:8080/

Note that the development build is not optimized.
To create a production build, run npm run build.
```

Figure 8.1: Output from running npx

You can access the application using the URL specified in the output – for example, `http://localhost:8080`. *Figure 8.2* shows the output of the application:

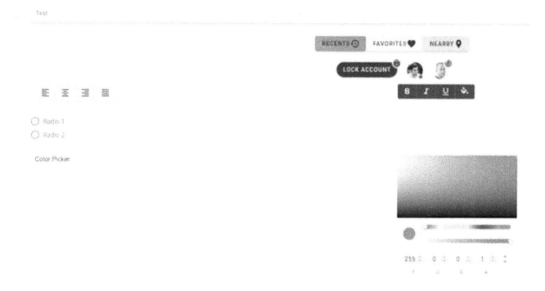

Figure 8.2: Output from the sample app

The sample app displays the different components that are available inside Vuetify. As you can see, there are components available for radio button groups and color pickers, among many others.

In the next section, we will look at how we use Vuetify in the sample app and how things are wired up together.

Using UI components

There are many components provided with Vuetify but in this section, we will just talk about a few of them to get an idea of how to use them. The example code uses components such as a color picker, button, badge, and so on.

Figure 8.3 shows the directory structure of the example. All of the source files are inside the src/ folder:

Figure 8.3: The directory structure of a Vuetify sample app

The main.js host code for initializing Vue and Vuetify is shown in the following snippet:

```
import Vue from 'vue'
import App from './App.vue'
import vuetify from './plugins/vuetify';

Vue.config.productionTip = false

new Vue({
 vuetify,
 render: h => h(App)
}).$mount('#app')
```

The code looks like any other Vue-based application except it adds the Vuetify framework, which is imported from the plugins/vuetify directory, as shown in this snippet:

```
import Vue from 'vue';
import Vuetify from 'vuetify/lib/framework';

Vue.use(Vuetify);
export default new Vuetify({});
```

Vuetify is initialized in the code as a plugin using the `Vue.use()` function call and exported to be made available to other parts of the code.

Now that the initialization is out of the way, let's take a look at how the sample is using the Vuetify components. The code snippet here from `App.vue` shows how the sample code uses the Color Picker component of Vuetify:

```
<template>
  <v-app>
    <v-container>
      . . .
      <v-row>
        <v-col>
          Color Picker
        </v-col>
        <v-col>
          <v-color-picker/>
        </v-col>
      </v-row>
    </v-container>
  </v-app>
</template>
```

The tags that can be seen in the snippet – `<v-row>`, `<v-col>`, `<v-container>`, and so on – are all Vuetify components. The components can be configured through the available properties; for example, if we look at the component documentation (`https://vuetifyjs.com/en/api/v-row/#props`) for `<v-row>`, we can see that we can set different parameters, such as alignment.

In this section, we learned about Vuetify and how to use the components provided, and also how to wire things together to use it in a Vue-based application. In the next section, we will look at different user interface libraries that are more lightweight compared to Vuetify. We will start by looking at Buefy in the next section.

Understanding Buefy

Buefy is another user interface framework that is built on top of Bulma. Bulma (`https://bulma.io/`) is an open source CSS project that provides different kinds of styles for HTML elements; the CSS file can be viewed at the following link: `https://github.com/jgthms/bulma/blob/master/css/bulma.css`.

Let's take a quick look at an example web page that uses Bulma CSS. This will give us a better idea of what Bulma is all about and also give us a better understanding of how Buefy is using it.

Bulma sample

Open the sample `chapter08/bulma/bulma_sample.html` file in your browser, and the HTML page will look like *Figure 8.4*:

Figure 8.4: Bulma example page

The following code snippet shows the Bulma CSS file used in the web page:

```
<head>
   ...
   <link rel="stylesheet" href=
      "https://cdn.jsdelivr.net/npm/bulma@0.9.3/css/
      bulma.min.css">
</head>
```

The web page uses different HTML elements tags styled using the Bulma CSS, as seen in the following code snippet:

```
<section class="hero is-medium is-primary">
    <div class="hero-body">
        <div class="container">
            <div class="columns">
                . . .
            </div>
        </div>
    </div>
</section>
<section class="section">
    <div class="container">
        <div class="columns">
            <div class="column is-8-desktop
                        is-offset-2-desktop">
                <div class="content">
                    . . .
                </div>
            </div>
        </div>
    </div>
</section>
```

Now that we have an idea about what Bulma is and how to use it for a web page, we will take a look at setting up Buefy in the next section.

Setting up Buefy

We are going to look at the Buefy example that is found inside the `chapter8/buefy` directory. Make sure you are inside the directory and run the following command:

```
npm install
npx vue-cli-service serve
```

Open the server in your browser and you will see output like *Figure 8.5*:

Figure 8.5: Buefy sample output

UI components

The web page displays different components available in Buefy, such as a slider, a clickable button with a dropdown, and a breadcrumb.

Initializing Buefy is the same as initializing any other Vue plugin. It looks the same as what we went through in the previous section when we looked at Vuetify. The code will initialize Vue by using Buefy as stated in the Vue.use(Buefy) code:

```
import Vue from 'vue'
import App from './App.vue'
import Buefy from "buefy";

Vue.use(Buefy);

new Vue({
  render: h => h(App)
}).$mount('#app')
```

One of the components that we are using in our sample app is `carousel`, which displays a user interface like a slideshow. To create `carousel`, it is just a few lines of code, as shown in the following code snippet, using the `<b-carousel>` tag:

```
<!--example from https://buefy.org/documentation-->
<template>
 <section>
   <div class="container">
     <b-carousel>
       <b-carousel-item v-for="(carousel, i) in carousels"
        :key="i">
         <section :class="`hero is-medium
                           is-${carousel.color}`">
           <div class="hero-body has-text-centered">
             <h1 class="title">{{ carousel.text }}</h1>
           </div>
         </section>
       </b-carousel-item>
     </b-carousel>
   </div>
 ...

  </section>
</template>
```

Like `carousel`, there are many different pre-built components available in Buefy that can help design complex user interfaces.

In the next section, we will look at how we can use the Vuelidate library as a way to perform validation on the data we capture and present in our user interface to ensure we interpret our customers' data correctly.

Validating data entry with Vuelidate

If your app does anything interactive, it's likely that it will handle user-entered data, which means you must check whether what the users are providing is valid input.

Input validation libraries can be used to ensure only valid data is entered by the user and provide feedback as soon as data is received. This means we're validating as soon our user hits that input field!

We're going to explore HTML form validation in the frontend and the difference between input and value validation. It's also important to note that no matter the validation in the frontend, it's no substitute for validation in the backend and of the API endpoints. Our goal in the frontend is to prevent the user from making errors; however, you'll never stop bad guys from submitting bad data to your app.

We can look at frontend validation through two lenses, as there's a myriad of solutions out there, but we'll contrast two options and show a working solution – the first is that of validating input, and the other is the validation of values.

If we only want to validate the input, we could use the `vee-validate` library, which works by having you write the rules inside the `<template>` of your code. For example, see the following:

```
<script>
Vue.use(VeeValidate);

var app = new Vue({
  el: '#app',
  data: {
    email: '',
  },
  methods: {
    onSubmit: function(scope) {
      this.errors.clear(scope);
      this.$validator.validateAll(scope);
    }
  }
});
</script>

<template>
<div>
  <form v-on:submit.prevent="onSubmit('scope')">
    <div>
      <div v-for="error in errors.all('scope')">
        {{error}}
      </div>
    </div>
    <div>
      <label>Email Address</label>
```

```
      <input type="text" v-model="email"
        name="Email Address" v-validate data-scope="scope"
        data-rules="required|min:6|email">
    </div>
    <div>
      <button type="submit">
        Send
      </button>
    </div>
  </form>

  <div class="debug">
    email: {{email}}<br>
  </div>
 </div>
</template>
```

This inline validation – wherein we perform `ValidateAll()` on submitting data – will allow us to validate the contents of the data using predefined rules, such as a field being required, its minimum length, or that it must be a valid email ID, for example. If invalid data is entered, we can iterate through the errors and present them to the user:

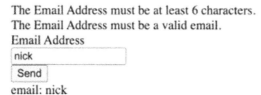

Figure 8.6: Validation error message

You can see this on the JS Playground website `JSFiddle` at the following link: `https://jsfiddle.net/vteudms5/`.

This is useful for simple validation, but when we want to add additional logic against values and collections of values, rather than just individual inputs, this is where libraries such as Vuelidate become powerful.

With Vuelidate, you'll notice that the validation is decoupled from the template code we write, unlike the inline validation done in the `vee-validate` example. This allows us to write the rules against the data model rather than the inputs in our template.

In Vuelidate, the validation results in a validation object referenced as `this.$v`, which we can use to validate our model state. Let's rebuild our previous example to demonstrate how we going to use Vuelidate to validate the data – this sample is at `https://jsfiddle.net/34gr7vq0/3/`:

```
<script>
Vue.use(window.vuelidate.default)
const { required, minLength, email } = window.validators

new Vue({
    el: "#app",
  data: {
      text: ''
  },
  validations: {
      text: {
         required,
         email,
         minLength: minLength(2)
      }
  },
  methods: {
      status(validation) {
         return {
         error: validation.$error,
         dirty: validation.$dirty
      }
    }
  }
})
</script>

<template>
<div>
  <form>
    <div>
       <label>Email Address</label>
       <input v-model="$v.text.$model"
```

```
        :class="status($v.text)">
      <pre>{{ $v }}</pre>
    <div>
  </form>
</div>
</template>
```

The resulting output shows us the $v object. The `required`, `email`, and `minLength` fields are firing when you type in the box. In our example, when we type in `nick@bar.com`, the fields change value:

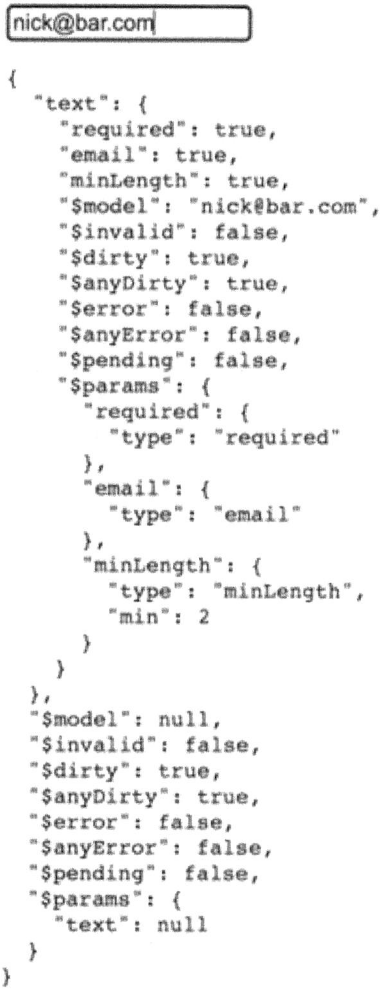

```
nick@bar.com

{
  "text": {
    "required": true,
    "email": true,
    "minLength": true,
    "$model": "nick@bar.com",
    "$invalid": false,
    "$dirty": true,
    "$anyDirty": true,
    "$error": false,
    "$anyError": false,
    "$pending": false,
    "$params": {
      "required": {
        "type": "required"
      },
      "email": {
        "type": "email"
      },
      "minLength": {
        "type": "minLength",
        "min": 2
      }
    }
  },
  "$model": null,
  "$invalid": false,
  "$dirty": true,
  "$anyDirty": true,
  "$error": false,
  "$anyError": false,
  "$pending": false,
  "$params": {
    "text": null
  }
}
```

Figure 8.7: Illustration of the browser output from our JSFiddle sample

Although similar to the `vee-validate` implementation in style, by utilizing the `$v` object concept and allowing that to be the source of validation, we can connect it to additional inputs across multiple forms and validate the entire collection. For example, if we had multiple fields, such as a name, email, users, and tags across `formA` and `formB`, we would be able to create the validation as follows:

```
...
validations: {
  name: { alpha },
  email: { required, email }
  users: {
    minLength: minLength(2)
  },
  tags: {
    maxLength: maxLength(5)
  },
  formA: ['name', 'email'],
  formB: ['users', 'tags']
}
```

There's a large collection of available validators for Vuelidate that we can import. This gives us access to validators such as conditionally required fields; length validators; email, alpha/alphanum, regex, decimal, integer, and URL options; and many more that are accessible by importing the `validators` library:

```
import { required, maxLength, email } from '@vuelidate/
validators'
```

The full list is available on the Vuelidate website at `https://vuelidate-next.netlify.app/validators.html`.

Better input handling with Cleave.JS

As we've just seen, getting data from your users in the right shape and form can be a challenge – be it a date in a YYYY/MM format, a phone number with a prefix, or other more structured input types.

We looked at validation previously, but you can further help your users by providing visual clues and feedback as they type to prevent them from reaching the end with validation errors – libraries such as those provided by the popular credit card and online payments processor. Stripe does a great job at helping users enter their credit card info correctly, but for those of us on a budget, we can use Cleave. js for a similar experience.

Figure 8.7: Credit card validation (image from https://nosir.github.io/cleave.js/)

Frustratingly, Vue isn't supported as a first-class citizen but there's no reason we can't set up the directive, which is available at codesandbox.io here – https://bit.ly/3Ntvv27. *Figure 8.8* shows how the validation will work for codesandbox.io:

```
4242 4222 4242 4244
```
4242 4222 4242 4244 visa

Figure 8.8: Example of our Cleave.js example on codesandbox.io

It's not as pretty in my hardcoded sample (the CSS is left as an exercise for you!) but the key part from the sandbox sample is how we overload custom-input with our cleave directive by doing the following:

```
<template>
<div id="app">
  <div>
   <custom-input
    v-cleave="{ creditCard: true,
    onCreditCardTypeChanged: cardChanged, }"
    v-model="ccNumber" />
</div>
<pre>
{{ ccNumber }}
{{ cardType }}
</pre>
</template>
```

In the future, it would be great to see Cleave.js incorporate a first-party implementation for Vue but until then, a number of npm packages exist to skip over the setup for our sample and provide a similar effect that will allow us to create beautiful experiences for our users.

To follow the status of Cleave.js official support, you can check out `https://github.com/nosir/cleave.js/blob/master/doc/vue.md`.

With Cleave.js, we have reached the end of this chapter.

Summary

In this chapter, we learned about several frontend libraries and tools to help us to iterate through code and design faster when building the frontend user interface for our product.

We've looked at using Vuetify to create customizable user interfaces, and looked at Buefy, which provides a huge collection of UI components to allow us to build our apps rapidly.

We then finished up by providing an introduction to and contrast between input and value validation using Vuelidate and VeeValidate, respectively, and finally, explained how we can use Cleave.js to create smarter interfaces to help our users understand what our app expects.

In the next chapter, we will look at middleware pieces that will form the bridge between the frontend and the backend.

Tailwind, Middleware, and CORS

In this chapter, we will build on the frontend principles we introduced previously by introducing Tailwind CSS, explore how we can consume our backend services via an API from our frontend app, see how we can leverage middleware to transform our JSON requests, and look at how we can provide a secure **Single-Page App (SPA)** with a user login.

In this chapter, we'll cover the following topics:

- Creating and designing frontend applications using the Tailwind CSS framework
- Getting an understanding of how to use the Vite CLI to create new Vue applications
- Configuring our Go service for CORS
- Setting up a JavaScript Axios library
- Creating middleware to manage JSON formatting between the frontend and the backend

Technical requirements

All the source code explained in this chapter can be checked out at `https://github.com/PacktPublishing/Full-Stack-Web-Development-with-Go/tree/main/Chapter09`.

Introducing Tailwind

In the previous chapter, we looked at a number of different frontend frameworks to help us go faster, but we've been ignoring an elephant in the room of a modern web ecosystem – Tailwind CSS.

Frameworks such as Buefy and Vuetify have a major disadvantage. Due to increasing demand for more and more features, growth, and usage, they became a victim of their own success and ended up too big, giving us less control over our component styles.

Learning about frameworks such as Buefy has become increasingly challenging. Developers have to learn about hundreds of classes and components and then potentially rebuild them just for small style tweaks that were simply not envisioned by the upstream community.

Tailwind is a CSS framework that, unlike other frameworks, does not come prebuilt with classes to add to HTML tags. Instead, it uses a different approach. It brings a much lower level of control by removing ALL default styling from the stylesheet and using utility-based classes to compose and build your app. These utility-based classes provide ways to directly manipulate certain CSS attributes individually, such as text size, margins, spacing, padding, and colors, as well as behavior for mobile, desktop, and other viewports. By applying different tailwind modifiers, we have granular control over the final appearance of an element while ensuring consistent styling and an easy escape route if we need to build slight variations. This really helps in building our Vue components.

Figure 9.1: A button sample

A quick example of creating a blue button can be seen with the following:

```
<button type="button" class="
  inline-block px-6 py-2.5 bg-blue-600
  text-white font-medium text-lg leading-tight
  rounded shadow-md
  hover:bg-blue-700 hover:shadow-lg
  focus:bg-blue-700 focus:shadow-lg
  focus:outline-none focus:ring-0
  active:bg-blue-800 active:shadow-lg
  transition duration-150 ease-in-out
">Button</button>
```

You may be saying to yourself, "Wow, that's a lot of CSS for a button," but when you consider how Vue helps us build reusable **Single-File Components (SFCs)**, we would only need to style this once, and all of our components would share that same utility-based CSS approach – whether it's a button, link, image, div, or paragraph. You can check the official docs at https://tailwindcss. com/docs/utility-first to dive further into the concepts behind "utility-first" CSS and what the individual classes do.

Creating a new Tailwind and Vite project

To create our project, we're going to first generate it with the Vite CLI. This will give us the familiar "Hello World" output you see here:

Hello Vue 3 + Vite

Recommended IDE setup: VSCode + Volar

Vite Documentation | Vue 3 Documentation

count is: 0

Edit components/HelloWorld.vue to test hot module replacement

Figure 9.2: Hello World web output

Let's create a new Vue project with Vite using the following command:

```
npm create vite@latest
```

For each of the questions asked, enter the information shown here:

```
✓ Project name: … vue-frontend
✓ Select a framework: > vue
✓ Select a variant: > vue

Scaffolding project in /Users/.../vue-frontend...

Done. Now run:
```

```
cd vue-frontend
npm install
npm run dev


$ npm install
$ npm run dev


> vue-frontend@0.0.0 dev
> vite



vite v2.9.12 dev server running at:

> Local: http://localhost:3000/
> Network: use `--host` to expose

ready in 332ms.
```

Going to `http://localhost:3000` will now show the screenshot from *Figure 9.2*. Our project is enabled with "hot reload" or "live reload" so as you change the project code, you will be able to see the design in the browser update when you save the file.

Previous versions of Tailwind CSS had a bit of a reputation for generating large stylesheets (between 3 and 15 MB!) and slowing down build times.

At the end of the Tailwind CSS version 2 era, the team introduced a new **Just-In-Time (JIT)** compiler that automatically generates only the necessary CSS required to style your design. This was originally available as an optional plugin but brought massive improvements by reducing bloat, and with JIT, the CSS in development is the same as your final code, which meant no post-processing of the CSS is required for your final builds. Since Tailwind CSS version 3 and above, the JIT compiler has been enabled by default when we install Tailwind CSS, so we don't have to worry about altering anything in our config file other than what is needed to lay out our project.

We're going to now add Tailwind CSS to our project and make some changes to the default Vue `Hello World` output provided by the scaffolding from both the Vue and Tailwind packages:

```
$ npm install -D tailwindcss postcss autoprefixer
$ npx tailwindcss init -p


Created Tailwind CSS config file: tailwind.config.js
```

```
Created PostCSS config file: postcss.config.js

$ cat << EOF > tailwind.config.js
/** @type {import('tailwindcss').Config} */
module.exports = {
  content: [
    "./index.html",
    "./src/**/*.{vue,js}",
  ],
  theme: {
    extend: {},
  },
  plugins: [],
}

EOF
$ cat << EOF > ./src/tailwind.css
@tailwind base;
@tailwind components;
@tailwind utilities;

EOF
$ cat << EOF > ./src/main.js
import { createApp } from 'vue'
import App from './App.vue'
import './tailwind.css'

createApp(App).mount('#app')

EOF
```

The directives beginning with `@tailwind` in the `tailwind.css` file are part of how we tell the JIT compiler what to apply to generate the CSS – we will only leverage the base, component and utility directives and refer you to the Tailwind CSS official docs for more on this – `https://tailwindcss.com/docs/functions-and-directives`.

We can now open up our `HelloWorld.vue` file and replace the contents with the following to create our button. The cool part with our dev server still running is that you should be able to see the changes in real time if you save your file as you manipulate the `button` classes:

```
<template>
  <div class="flex space-x-2 justify-center">
    <button
      @click="count++"
      type="button"
      class="inline-block px-6 py-2.5 bg-blue-600
             text-white font-medium text-lg leading-tight
             normal-case rounded shadow-md hover:bg-blue-
             700 hover:shadow-lg focus:bg-blue-700
             focus:shadow-lg focus:outline-none
             focus:ring-0 active:bg-blue-800
             active:shadow-lg transition duration-150
             ease-in-out"
    >
      Click me - my count is {{ count }}
    </button>
  </div>
</template>
```

You should end up with something like this:

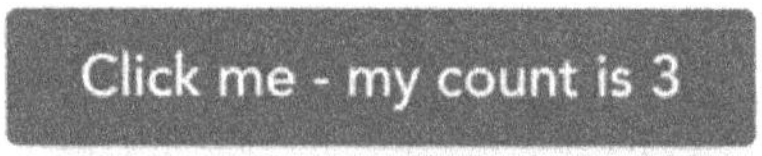

Figure 9.3: The Click me button

Congratulations! You've created your first Tailwind and Vite project. You can see the complete example inside the `chapter9/tailwind-vite-demo` folder.

In the next section, we will look at how to use the API that we developed in Golang from our frontend.

Consuming your Golang APIs

We're going to build on our previous frontend example to add some functions to GET and POST from a simple backend service. The source code can be found inside the `chapter9/backend` folder; it focuses on two simplified functions that do little more than return a fixed string for GET and a reversed string based on the POST request that we sent.

The `appGET()` function provides the functionality to perform a `GET` operation, while the `appPOST()` function provides it for a `POST` operation:

```go
func appGET() http.HandlerFunc {
    type ResponseBody struct {
        Message string
    }
    return func(rw http.ResponseWriter, req *http.Request) {
        log.Println("GET", req)
        json.NewEncoder(rw).Encode(ResponseBody{
            Message: "Hello World",
        })
    }
}

func appPOST() http.HandlerFunc {
    type RequestBody struct {
        Inbound string
    }
    type ResponseBody struct {
        OutBound string
    }
    return func(rw http.ResponseWriter, req *http.Request) {
        log.Println("POST", req)

        var rb RequestBody
        if err := json.NewDecoder(req.Body).Decode(&rb);
                err != nil {
            log.Println("apiAdminPatchUser: Decode
                    failed:", err)
            rw.WriteHeader(http.StatusBadRequest)
            return
        }
        log.Println("We received an inbound value of",
                rb.Inbound)
        json.NewEncoder(rw).Encode(ResponseBody{
            OutBound: stringutil.Reverse(rb.Inbound),
```

```
        })
      }
   }
```

We'll start our service by using go run server.go, with a view to consuming this data from our frontend application.

We're going to create two utility functions in our frontend app to allow us to interact with our frontend app, and we're going to be building these on top of Axios. Axios is a Promise-based HTTP client for the browser that abstracts all the browser-specific code needed to interact with backend services and does an incredible job in providing a single interface for web requests across all browsers , which you can read more about at the official docs here: https://axios-http.com/.

We're going to first install axios, then set up our Axios instance, and then we can layer on functionality:

```
$ npm install axios
```

With axios installed, you'll now want to create a lib/api.js file containing the following:

```
import axios from 'axios';

// Create our "axios" object and export
// to the general namespace. This lets us call it as
// api.post(), api.get() etc
export default axios.create({
  baseURL: import.meta.env.VITE_BASE_API_URL,
  withCredentials: true,
});
```

There's a couple of interesting things to note here; the first is the baseURL value, and the second is withCredentials.

The baseURL value is what Axios uses to build all subsequent requests on top of. If we called axios.Patch('/foo') with a baseURL value of https://www.packtpub.com/, it would perform a PATCH call to https://www.packtpub.com/foo. This is a great way to switch between development and production and ensure you reduce typos.

But what are we doing with import.meta.env? This is partly how Vite imports and exposes environment variables. We're going to add our VITE_BASE_API_URL to a .env file situated at the base of our project containing the following:

```
VITE_BASE_API_URL="http://0.0.0.0:8000"
```

Combined with this and our new `lib/api.js` file, we can now call `axios.Put('/test')` from our code, and by default, it will reference `http://0.0.0.0:8000/test`. You can see more about how Vite handles environment variables and more at `https://vitejs.dev/guide/env-and-mode.html`.

The other part to note is the `withCredentials` property. This value indicates whether or not cross-site access control requests should be made using credentials such as cookies and authorization headers.

The reason we want this property is that we want all our cookie settings to be consistent, but we'll need to ensure our backend app understands it, which we'll cover shortly. Setting `withCredentials` has no effect on same-site requests.

Now that we've used this to instantiate our `axios` instance, we can leverage this by creating our own `api/demo.js` file inside our frontend application's `src` folder. It's not a very original name but it works for us:

```
import api from '@/lib/api';



export function getFromServer() {
    return api.get(`/`);
}


export function postToServer(data) {
    return api.post(`/`, data );
}
```

This code exports two functions called `getFromServer` and `postToServer`, with an additional `data` parameter being sent as the `POST` body on the latter function.

A neat trick here is the usage of the @ import – this is common in a lot of setups to allow us to quickly specify the base path for our code to keep things clean and remove relative/absolute pathing with lots of `../..` referenced everywhere. If you forget this, you'll see errors such as this:

```
12:23:46 [vite] Internal server error: Failed to resolve import
"@/api/demo" from "src/components/HelloWorld.vue". Does the
file exist?
  Plugin: vite:import-analysis
  File: /Users/nickglynn/Projects/Becoming-a-Full-Stack-Go-
        Developer/chapter 9/frontend/src/components/
        HelloWorld.vue
  1  |  import { ref } from 'vue';
```

```
2 |    import * as demoAPI from '@/api/demo';
  |                            ^
3 |
4 |    // Sample to show how we can inspect mode
```

Not great! To fix this, open up your `vite.config.js` file and replace the contents with the following:

```
import { defineConfig } from 'vite'
import vue from '@vitejs/plugin-vue'
import path from 'path';


// https://vitejs.dev/config/
export default defineConfig({
  plugins: [vue()],
  // Add the '@' resolver
  resolve: {
    alias: {
      '@': path.resolve(__dirname, 'src'),
    },
  },
})
```

I've bolded the key parts that we're adding. We're telling Vite to use the @ symbol as an alias so that when we use @ in a path, it calls `path.resolve()` to resolve the path segments into an absolute path.

With all of this now set up, we're going to open up our `HelloWorld.vue` and update it, the goal being to create something that looks like *Figure 9.4*

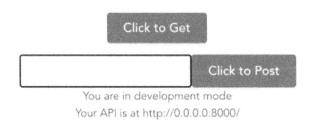

Figure 9.4: The UI for GET and POST

Here is the full code for `HelloWorld.vue`:

```
<script setup>
import { ref } from 'vue';
import * as demoAPI from '@/api/demo';

// Sample to show how we can inspect mode
// and import env variables
const deploymentMode = import.meta.env.MODE;
const myBaseURL = import.meta.env.VITE_BASE_API_URL;

async function getData() {
  const { data } = await demoAPI.getFromServer()
  result.value.push(data.Message)
}

async function postData() {
  const { data } = await demoAPI.postToServer({ Inbound: msg.
    value })
  result.value.push(data.OutBound)
}

const result = ref([])
const msg = ref("")

defineProps({
  sampleProp: String,
});

</script>

<template>
  <div class="flex space-2 justify-center">
    <button
      @click="getData()"
      type="button"
      class="inline-block px-6 py-2.5 bg-blue-600
```

```
                    text-white font-medium text-lg leading-tight
                    normal-case rounded shadow-md hover:bg-blue-
                    700 hover:shadow-lg focus:bg-blue-700
                    focus:shadow-lg focus:outline-none
                    focus:ring-0 active:bg-blue-800
                    active:shadow-lg transition
                    duration-150 ease-in-out"
      >
        Click to Get
      </button>
  </div>
  <div class="flex mt-4 space-2 justify-center">
    <input type="text"
      class="inline-block px-6 py-2.5 text-blue-600
             font-medium text-lg leading-tight
             rounded shadow-md border-2 border-solid
             border-black focus:shadow-lg  focus:ring-1 "
      v-model="msg" />
    <button
      @click="postData()"
      type="button"
      class="inline-block px-6 py-2.5 bg-blue-600
             text-white font-medium text-lg leading-tight
             normal-case rounded shadow-md hover:bg-blue-
             700 hover:shadow-lg focus:bg-blue-700
             focus:shadow-lg focus:outline-none
             focus:ring-0 active:bg-blue-800
             active:shadow-lg transition
             duration-150 ease-in-out"
      >
        Click to Post
      </button>
  </div>
  <p>You are in {{ deploymentMode }} mode</p>
  <p>Your API is at {{ myBaseURL }}</p>
  <li v-for="(r, index) in result">
```

```
    {{ r }}
  </li>

</template>

<style scoped></style>
```

The parts in bold are the most interesting parts. These show how we can use GET and POST with our data, using our libraries and API calls from the backend server that we set up, as well as how we can bind the data and reference it in our Vue modules.

Hopefully, after making all these changes, your Vite instance is still running; if not, start it with npm run dev, and you should get the screenshot from *Figure 9.4*. Click the **Click to Get** button and enter some data to send via the **Click to post** button.

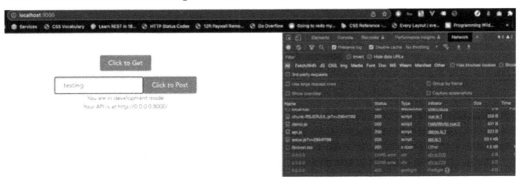

Figure 9.5: Peeking into the HTTP traffic

It doesn't work! We're so close, but first, we have to revisit CORS from one of our previous chapters.

CORS for secure applications

In *Chapter 6, Moving to API-First*, we introduced the CORS middleware for our backend. We've now got to update our new backend service. It will need to respond to OPTION preflight requests, as we discussed in *Chapter 6, Moving to API-First*, and will also need to identify the URLs that we're going to allow to talk to our service. This is necessary to ensure our browsers aren't being tricked into submitting/modifying applications from other sources.

Open up the backend/server.go sample you've been running and review the main function:

```
. . .
```

```
port := ":8000"
rtr := mux.NewRouter()
rtr.Handle("/", appGET()).Methods(http.MethodGet)
rtr.Handle("/", appPOST()).Methods(http.MethodPost,
                            http.MethodOptions)

// Apply the CORS middleware to our top-level router, with
// the defaults.
rtr.Use(
    handlers.CORS(
      handlers.AllowedHeaders(
        []string{"X-Requested-With",
        "Origin", "Content-Type",}),
      handlers.AllowedOrigins([]string{
        "http://0.0.0.0:3000",
        "http://localhost:3000"}),
        handlers.AllowCredentials(),
        handlers.AllowedMethods([]string{
            http.MethodGet,
            http.MethodPost,
        })),
    )

    log.Printf("Listening on http://0.0.0.0%s/", port)
    http.ListenAndServe(port, rtr)
```

As before, I've put the key parts in bold. You can see we've appended `http.MethodOptions` to our POST handler, and we've also layered in some additional middleware.

`AllowedHeaders` has been included, and we're specifically accepting `Content-Type` as, by default, we won't accept JSON – only `application/x-www-form-urlencoded`, `multipart/form-data`, or `text/plain` are accepted.

We also use `AllowCredentials` to specify that the user agent may pass authentication details along with the request, and finally, we're specifying our dev server's location, both for `localhost` and the `0.0.0.0` address. This might be slight overkill but can help if your backend and frontend start differently.

For a production-ready version of our project, you will want to inject these as environment variables to avoid mixing development and production config files. If you leverage `env.go` from

Chapter 6, Moving to API - First – available at `https://github.com/PacktPublishing/`
`Full-Stack-Web-Development-with-Go/blob/main/Chapter06/internal/env.`
`go` – you will do something like the following:

```
rtr.Use(
    handlers.CORS(
        handlers.AllowedHeaders(
            env.GetAsSlice("ALLOWED_HEADERS")),
        handlers.AllowedOrigins(
            env.GetAsSlice("ORIGIN_WHITELIST")),
        handlers.AllowCredentials(),
        handlers.AllowedMethods([]string{
            http.MethodGet,
            http.MethodPost,
        })),
)
```

Once your server is configured correctly, (re)start both the backend and the frontend, and you should now be able to call your backend service to use GET and POST. You've now completed a full-stack project!

Figure 9.6: UI displaying output from the server

In this section, we looked at adding CORS functionality to our application, allowing the frontend to access our API. In the next section, we will look at exploring Vue middleware that will help to provide common data transformation functionality.

Creating Vue middleware

Working with Vue (and Axios) and Golang, we've shown we can bring all of learning so far all together, but we've missed one small aspect. We've deliberately omitted the JSON `struct` tags from our Golang code. If we add them back into our `backend/server.go` and rerun both the server and app, our requests no longer work!

```
func appPOST() http.HandlerFunc {
    type RequestBody struct {
        InboundMsg string `json:"inbound_msg,omitempty"`
    }
    type ResponseBody struct {
        OutboundMsg string `json:"outbound_msg,omitempty"`
    }
...
```

Our frontend and backend can no longer communicate as the contract has changed; the frontend is communicating in CamelCase, while the backend is communicating in snake_case.

This isn't a show-stopper, as we've proven we can work around it, but sometimes we don't have the luxury of telling the backend service what format to use. Thankfully, Axios can be modified to add transformers to our requests that will modify inbound and outbound requests to match whichever backend formatting we're given.

To build our transformers, we'll install and use two new packages to help us to create our transformers. These will be used to convert between the different formats/case types:

```
$ npm install snakecase-keys camelcase-keys
```

Finally, we'll modify our `lib/api.js` file to use these libraries to format our payloads:

```
import axios from 'axios';
import camelCaseKeys from 'camelcase-keys';
import snakeCaseKeys from 'snakecase-keys';


function isObject(value) {
  return typeof value === 'object' && value instanceof
    Object;
}
```

```
export function transformSnakeCase(data) {
  if (isObject(data) || Array.isArray(data)) {
    return snakeCaseKeys(data, { deep: true });
  }
  if (typeof data === 'string') {
    try {
      const parsedString = JSON.parse(data);
      const snakeCase = snakeCaseKeys(parsedString, { deep:
                                      true });
      return JSON.stringify(snakeCase);
    } catch (error) {
      // Bailout with no modification
      return data;
    }
  }
  return data;
}


export function transformCamelCase(data) {
  if (isObject(data) || Array.isArray(data)) {
    return camelCaseKeys(data, { deep: true });
  }
  return data;
}


export default axios.create({
  baseURL: import.meta.env.VITE_BASE_API_URL,
  withCredentials: true,
  transformRequest: [...axios.defaults.transformRequest,
                     transformSnakeCase],
  transformResponse: [...axios.defaults.transformResponse,
                     transformCamelCase],
});
```

This code might look like a lot, but it's what we need to create our transformers. We create a `to` function and a `from` function to add as transformers to the Axios instantiation. We transform the requests into snake_case on the outbound/request and transform them to CamelCase on the inbound/response. If you want to dive into the specifics of creating transformers for Axios, you can find more on the website at `https://axios-http.com/docs/req_config`, which includes a look at all the other numerous configs and parameters that can be provided for the Axios library.

There are a few different methods/libraries we could use to accomplish the same goal – for example, the `humps` package from `https://www.npmjs.com/package/humps` is another library we could use to expose similar functionality, but what we are using works well for our use case.

Summary

This chapter introduced Tailwind CSS and discussed its utility-first approach. We've previously seen samples of it in *Chapter 4, Serving and Embedding HTML Content*, where we were provided with the HTML/CSS, but this is our first look at using it and how we can rapidly create components outside of heavier frameworks, as well as how we can rapidly integrate it with our frontend Vue application with configuration and how we can test its successful installation.

In this chapter, we created a full-stack application, bringing our expertise together thus far. We've successfully built a frontend application in Vue that communicates with our backend in Golang; as part of this, we also looked at how to configure and use Axios and how to mitigate common CORS issues, before concluding with a brief look at using middleware in our Vue app to allow us to communicate across different JSON schemas in the backend.

In the next chapter, we'll look into securing our sessions, using JWTs for sessions, middleware, and creating and using navigation guards in Vue.

10

Session Management

In *Chapter 9, Tailwind, Middleware, and CORS*, we created a full-stack app with an independent frontend and backend talking to each other via an API.

In this chapter, we'll bring all of our existing knowledge together, introduce how to create and validate JSON Web Tokens (JWTs) for session management and middleware, set up the basic tenets of using Vue Router with navigation guards, and learn about errors and "catch-all" navigation guards.

We'll cover the following topics in this chapter:

- Session management and JWTs
- (Re)introducing Vue Router
- Navigation guards
- Defaults and error pages

By the end of this chapter, we'll have an understanding of how to complete and secure a project ready for our waiting users.

Technical requirements

All the source code explained in this chapter can be checked out at `https://github.com/PacktPublishing/Full-Stack-Web-Development-with-Go/tree/main/chapter10`.

Session management and JWTs

We looked at session management using cookies previously in *Chapter 6, Moving to API-First*, using the Gorilla Mux middleware. In our app, we created an in-memory cookie store via the functionality provided by Gorilla sessions: `https://github.com/gorilla/sessions`.

We previously implemented our middleware to validate that our user was approved by encoding two values – a user ID we looked up from the database and a `userAuthenticated` Boolean value.

This worked well for our use case, but our implementation meant that every call to our API backend required a round trip to the database to check that the user ID was still present, before letting the call continue.

Login

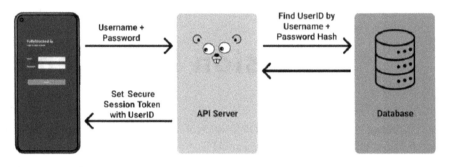

Save Workout

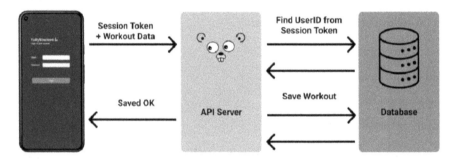

Figure 10.1: An illustration of login and save API workflows using a session cookie

This approach is fine and the Gorilla sessions library provides a number of alternative backends to speed things up, such as using Redis and SQLite, but we're going to look at an alternative approach using JWTs.

What's a JWT?

JWT stands for JSON Web Token. A JWT is a standard for creating data with optional signatures (public or public/private) and/or encryption, with a payload consisting of JSON that asserts a number of what the JWT specification calls claims. You can generate and examine JWTs on the web at jwt.io, and these are broken down into three parts, consisting of the header, the payload (with the claims), and the signature. These are then base64-encoded and concatenated together using a . separator, which you can see here.

eyJhbGciOiJIUzI1NiIsInR5cCI6IkpXVCJ9.ey
JzdWIiOiIxMjM0NTY3ODkwIiwidGl0bGUiOiJGd
WxsIFN0YWNrIEdvIiwiaWF0IjoxNTE2MjM5MDIy
fQ.8ph4Comr-law35-
d1kJSDS3riNiS7GsF2NwJNB8I-0M

**Header - algorithm used
and the token type**

**Payload - The data and
claims**

Signature - Used to
verify authenticity

Figure 10.2: Color-coded illustration showing the parts of a JWT

The part that is of interest to us is the payload and the claims. A number of reserved claims exist that we should respect as part of the specification, which are as follows:

- **iss** (**issuer**): The issuer of the JWT.

- **sub** (**subject**): The subject of the JWT (the user).

- **aud** (**audience**): The recipient for which the JWT is intended.

- **exp** (**expiration time**): The time after which the JWT expires.

- **nbf** (**not before time**): The time before which the JWT must not be accepted for processing.

- **iat** (**issued at time**): The time at which the JWT was issued. This can be used to determine the age of the JWT.

- **jti** (**JWT ID**): A unique identifier. This can be used to prevent the JWT from being replayed (allows a token to be used only once).

In the library, we're going to use `go-jwt`, available at `https://github.com/golang-jwt/jwt`. These standard claims are provided via a Go struct, as shown here:

```
// Structured version of Claims Section, as referenced at
// https://tools.ietf.org/html/rfc7519#section-4.1
type StandardClaims struct {
    Audience  string `json:"aud,omitempty"`
    ExpiresAt int64  `json:"exp,omitempty"`
    Id        string `json:"jti,omitempty"`
    IssuedAt  int64  `json:"iat,omitempty"`
    Issuer    string `json:"iss,omitempty"`
    NotBefore int64  `json:"nbf,omitempty"`
    Subject   string `json:"sub,omitempty"`
}
```

We can add to these claims to provide our own additional claims, and in typical Go style, we do so by embedding `StandardClaims` into our own struct, which I've called `MyCustomClaims`, as shown here:

```go
mySigningKey := []byte("PacktPub")

// Your claims above and beyond the default
type MyCustomClaims struct {
    Foo string `json:"foo"`
    jwt.StandardClaims
}

// Create the Claims
claims := MyCustomClaims{
    "bar",
    // Note we embed the standard claims here
    jwt.StandardClaims{
        ExpiresAt: time.Now().Add(time.Minute *
                                   1).Unix(),
        Issuer:    "FullStackGo",
    },
}

// Encode to token
token := jwt.NewWithClaims(jwt.SigningMethodHS256,
                           claims)
tokenString, err := token.SignedString(mySigningKey)
fmt.Printf("Your JWT as a string is %v\n", tokenString)
```

If you execute this code, you will get the following output:

```
$ go run chapter10/jwt-example.go
Your JWT as a string is eyJhbGciOiJIUzI1NiIsInR5cCI6IkpXVCJ9.ey
Jmb28iOiJiYXIiLCJleHAiOjE2NTY3MzY2NDIsImlzcyI6IkZ1bGxTdGFja0dv
In0.o4YUzyw1BUukYg5H6CP_nz9gAmI2AylvNXG0YC5OE0M
```

When you run the sample code or write your own, it will look slightly different because of the relative expiration in `StandardClaims`, and if you tried decoding the preceding string, chances are that it will show as expired by quite a few seconds!

You may be asking why you should care about JWTs when you've already seen your database-based middleware working. The reason is that we can save a round trip to our database, saving us time and bandwidth.

Because JWTs are signed, we can confidently assume that the provided claims can be asserted to be true so long as the JWT is decoded as we expect. With our JWT-based model, we can instead encode the user details and permissions into the claims on the JWT itself.

Login

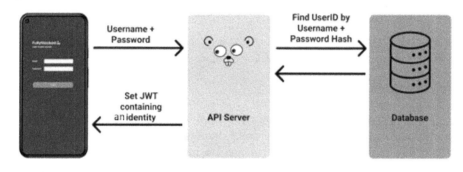

Save Workout

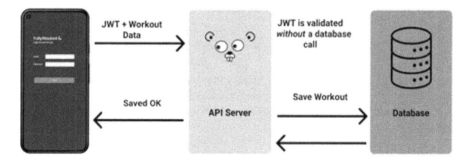

Figure 10.3: An illustration of login and save API workflows using a JWT secured session

This all seems great, but there are a number of "gotchas" when working with JWTs, and it's worth covering them before we start using them in every situation.

The "none algorithm" gotcha

An unsecured JWT can be created where the "`alg`" header parameter value is set to "`none`" with an empty string for its signature value.

Given that our JWTs are simply base64-encoded payloads, a malicious hacker could decode our JWT, strip off the signature, change the alg parameter to "none" and try to present it back to our API as a valid JWT.

```
$ Pipe our encoded JWT through the base64 command to decode it
$ echo eyJhbGciOiJIUzI1NiIsInR5cCI6IkpXVCJ9 | base64 -D
{"alg":"HS256","typ":"JWT"}
$ echo '{"alg":"none","typ":"JWT"}' | base64
eyJhbGciOiJub25lIiwidHlwIjoiSldUIn0K
```

It's important that the library you're using verifies that you're receiving your JWTs back with the same alg you provided, and you should verify this yourself before using it.

The "logout" gotcha

When you click to log out of your web app, the common thing to do is to set the cookie expiration to a date in the past, and then the browser will delete the cookie. You should also remove any active session information from your database and/or app. The issue is that with JWTs, it may not work how you expect it to. Because a JWT is self-contained, it will continue to work and be considered valid until it expires – the JWT expiration, not that of the cookie – so if someone were to intercept your JWT, they could continue to access the platform until the JWT expired.

The "banlist" or "stale data" gotcha

Similar to the logout gotcha, because our JWTs are self-contained, the data stored in them can be stale until refreshed. This can manifest as access rights/permissions becoming out of sync or, worse still, someone being able to continue to log in to your application after you've banned them. This is worse in scenarios where you need to be able to block a user in real time – for example, in situations of abuse or poor behavior. Instead, with the JWT model, the user will continue to have access until the token expires.

Using JWTs with cookies and our middleware

With all of our previous gotchas understood, we're going to write some simple middleware and cookie handling to build on our simple API service from Chapter 9, Tailwind, Middleware, and CORS, combining it with our knowledge from *Chapter 5, Securing the Backend and Middleware*.

This code is all provided on GitHub under `chapter10/simple-backend`.

Setting cookies and validation middleware

In order to start using our new JWTs, we're going to write some middleware for the mux to consume that we will inject into all our protected routes. As before, we're using a signature that the default library uses, where we take in http.Handler and return handlerFunc. When successful, we call next.ServerHTTP(http.ResponseWriter, *http.Request) to continue and indicate the successful handling of a request:

```
// JWTProtectedMiddleware verifies a valid JWT exists in
// our cookie and if not, encourages the consumer to login
// again.
func JWTProtectedMiddleware(next http.Handler) http.Handler {
    return http.HandlerFunc(func(w http.ResponseWriter,
                                 r *http.Request) {

        // Grab jwt-token cookie
        jwtCookie, err := r.Cookie("jwt-token")
        if err != nil {
            log.Println("Error occurred reading cookie", err)
            w.WriteHeader(http.StatusUnauthorized)
            json.NewEncoder(w).Encode(struct {
                Message string `json:"message,omitempty"`
            }{
                Message: "Your session is not valid -
                          please login",
            })
            return
        }

        // Decode and validate JWT if there is one
        userEmail, err := decodeJWTToUser(jwtCookie.Value)

        if userEmail == "" || err != nil {
            log.Println("Error decoding token", err)
            w.WriteHeader(http.StatusUnauthorized)
            json.NewEncoder(w).Encode(struct {
                Message string `json:"message,omitempty"`
            }{
```

```
                    Message: "Your session is not valid -
                          please login",
          })
          return
     }

     // If it's good, update the expiry time
     freshToken := createJWTTokenForUser(userEmail)

     // Set the new cookie and continue into the handler
     w.Header().Add("Content-Type", "application/json")
     http.SetCookie(w, authCookie(freshToken))
     next.ServeHTTP(w, r)
   })
}
```

This code is checking for our cookie, named jwt-token, and decodes it with our new decodeJWTToUser, checking the value for a valid entry. In our case, we expect userEmail, and if it is not present, we simply return an invalid session message. In this example, we then update the expiry time for the JWT and exit the function after setting the latest cookie.

In practice, we would check more strictly to ensure that a small window is kept for valid claims, and we'd then go back to the database to check whether the user still had permission to access our platform.

The functionality we use for setup and manipulation of our cookies is very similar to our previous work in *Chapter 5, Securing the Backend and Middleware* including with the domain, same-site mode, and, most importantly, HttpOnly and Secure.

We use Secure as good practice to ensure that it's only ever sent via secure HTTPS (except on localhost for development) as, although we can be confident our JWT is secure, it can still be decoded with tools such as jwt.io:

```
var jwtSigningKey []byte
var defaultCookie http.Cookie
var jwtSessionLength time.Duration
var jwtSigningMethod = jwt.SigningMethodHS256

func init() {
    jwtSigningKey = []byte(env.GetAsString(
```

```
            "JWT_SIGNING_KEY", "PacktPub"))
        defaultSecureCookie = http.Cookie{
            HttpOnly: true,
            SameSite: http.SameSiteLaxMode,
            Domain:   env.GetAsString("COOKIE_DOMAIN",
                                     "localhost"),
            Secure:   env.GetAsBool("COOKIE_SECURE", true),
        }
        jwtSessionLength = time.Duration(env.GetAsInt(
            "JWT_SESSION_LENGTH", 5))
}


...


func authCookie(token string) *http.Cookie {
    d := defaultSecureCookie
    d.Name = "jwt-token"
    d.Value = token
    d.Path = "/"
    return &d
}


func expiredAuthCookie() *http.Cookie {
    d := defaultSecureCookie
    d.Name = "jwt-token"
    d.Value = ""
    d.Path = "/"
    d.MaxAge = -1
    // set our expiration to some date in the distant
    // past
    d.Expires = time.Date(1983, 7, 26, 20, 34, 58,
                          651387237, time.UTC)
    return &d
}
```

The `HttpOnly` flag is used for us in our cookie package and hasn't been mentioned before – so, what is it?

Well, by default, when we don't use `HttpOnly`, our frontend Javascript can read and inspect cookie values. This is useful for setting a temporary state via the frontend and for storing a state that the frontend needs to manipulate. This is fine for a number of scenarios, and your application may have a combination of cookie-handling techniques.

When you use `HttpOnly`, the browser prevents access to the cookie, typically returning an empty string as the result of any values read. This is useful for preventing **Cross-Site Scripting** (**XSS**), where malicious sites try to access your values, and prevents you from sending data to a third-party/attacker's website.

This doesn't prevent us from logging in (which wouldn't be very helpful!). All our API/backend requests can still be performed with all cookies, but we do need to tell our frontend application to do so.

After providing the ability to log in with these additions to our backend, we're now going to revisit routing so that we can move around within our app.

(Re)introducing Vue Router

Before we dive in, we need to quickly refresh our understanding of how our frontend and backend communicate and ensure that we know how things work.

You may recall from *Chapter 9*, *Tailwind, Middleware, and CORS* that we set up our `axios` instance (under `src/lib/api.js`). With a few defaults, this is where the `withCredentials` value comes into play:

```
export default axios.create({
  baseURL: import.meta.env.VITE_BASE_API_URL,
  withCredentials: true,
  transformRequest: [...axios.defaults.transformRequest,
                     transformSnakeCase],
  transformResponse: [...axios.defaults.transformResponse,
                      transformCamelCase],
});
```

We want to ensure that all our hard work with the Secure and `HttpOnly` cookies is preserved when the frontend and backend communicate, and `withCredentials` ensures that all requests to the backend should be made, complete with cookies, auth headers, and so on.

We're going to be building on this `axios` instance as we introduce the concept of navigation guards. What we're going to do before we navigate around our application is fetch/refresh our data from the backend before rendering. This gives us the ability to check whether users should be looking at certain pages, whether they need to be logged in, or whether they shouldn't be snooping!

With our app now passing our cookies into every request, we can now move into utilizing permissions as we navigate our app using navigation guards.

Navigation guards

Navigation guards in Vue are fundamental for logged-in users. As with any core functionality of Vue, it's worth diving into the amazing documentation provided by the Vue team here: `https://router.vuejs.org/guide/advanced/navigation-guards.html`.

A navigation guard is, as the name suggests, a way to cancel or reroute users depending on the results of certain guard rails checks. They can be installed globally – for example, everything is behind a login/paywall – or they can be placed on individual routes.

They are called on a navigation request, in order, and before a component is loaded. They can also be used to retrieve props to be provided to the next pages components and use the syntax of `router.beforeEach(`**`to, from`**`)`.

Previous versions also provided a `next` parameter, but this has been deprecated and shouldn't be used in modern code.

The functionality of a navigation guard is as follows:

- `to`: Provides the target location, where the user is trying to navigate to

- `from`: The current location where the user is coming from

The job of the guard handler is to assess whether to allow navigation or not.

The handler can do this by returning `false`, a new route location, which is used to manipulate the browser history via a `router.push()` to allow additional props, or simply `true` to indicate the navigation is allowed.

Using a simple example from the docs, we can install a global navigation guard on our routes to check the value of the `isAuthenticated` variable before navigating:

```
router.beforeEach(async (to, from) => {
  if (
    // make sure the user is authenticated
    !isAuthenticated &&
    // Avoid an infinite redirect
    to.name !== 'Login'
  ) {
    // redirect the user to the login page
    return { name: 'Login' }
```

```
    }
  // Otherwise navigation succeeds to 'from'
})
```

Putting the logic into each route can be a bit ugly. What we will do is expose an endpoint in the backend that returns either a value or even just a 20x HTTP successful response, check for this in our middleware, and if that works, we will allow navigation.

In the following code, we've got an endpoint, /profile, exposed on our backend. This can return data or, in this simple case, just a 200/OK response, and we can check that with our getCheckLogin() function.

Our checkAuth() function now checks a meta value for an optional Boolean value called requiresAuth. If there's no authorization required, we navigate successfully; otherwise, we try to access our endpoint. If there's an error (non-successful) request, we redirect to our login page; otherwise, we allow the navigation to continue:

```
export function getCheckLogin() {
  return api.get('/profile');
}



export default function checkAuth() {

  return async function checkAuthOrRedirect(to, from) {
    if (!to?.meta?.requiresAuth) {
      // non protected route, allow it
      return;
    }
    try {
      const { data } = await getCheckLogin();
      return;
    } catch (error) {
      return { name: 'Login'};
    }
  };
}
```

These checks can be as complicated as we want in our navigation guards, but remember that you're calling these on every navigation. You might want to look at state management if you find yourself doing this a lot, such as Pinia (Vue 3) or Vuex (if you're using Vue 2).

To install these checks and values, we simply install the global handler, and for protected routes, we provide the `meta` Boolean. This is shown in the following code snippet:

```
...
const router = createRouter({
  history: createWebHistory(import.meta.env.BASE_URL),
  routes: [
{
    path: '/login',
    Name: 'Login',
    meta: {
      requiresAuth: false,
    },
    props: true,
    component: () => import('@/views/login.vue'),
  },{
    path: '/dashboard,
    Name: 'Dashboard',
    meta: {
      requiresAuth: true,
    },
    props: true,
    component: () => import('@/views/dashboard.vue'),
  }]
});
...
router.beforeEach(checkAuth());
```

Meta fields are a useful feature. They allow us to attach arbitrary information to our routes, in our situation we're using the meta information as an indicator to check the authorization. You can find out more about meta here: `https://v3.router.vuejs.org/guide/advanced/meta.html`.

With the ability to provide for logged-in and logged-out statuses, we now have a functioning app. One final thing to really polish our app is to provide default and error pages for our users if our app goes wrong or if they land on the wrong page in it.

Defaults and error pages

With our application now securely communicating to the backend and routing correctly based on authorization, we are almost finished with our core functional requirements.

There's one final scenario that may arise for our users – the dreaded 404 – the page not found error! Thankfully, Vue Router makes it easy to create a wildcarded "catch-all" route that will be set to redirect users to a specific page if they navigate to somewhere that doesn't exist.

As you know, in Vue, all routes are defined by creating rules on the specific URL path. So, for example, creating a route for a path of /user would be caught if the user entered packt.com/user, but it wouldn't if the user entered packt.com/my-user or any other word that is not precisely the one set in the path rule.

To define our catch-all rule in version 4 of the Vue routervue-router 4, we will use the following route entry:

```
{ path: '/:pathMatch(.*)*', name: 'not-found', component:
NotFound }
```

We will inject this as the final route in our router.routes. The wildcard at the end of the path match means we can navigate to this page and catch the expected route. Alternatively, if that's too much magic, you can use path: '/*' and don't need to worry about catching the intended route.

The best practice for a 404 page not found error would be to provide hints of what went wrong and give the user a way to get home or navigate to a similar page, but that's a choice you can make for your NotFound component.

Summary

Excitedly, we've now got enough knowledge to complete the development of our full-stack app. In this chapter, we introduced JWT-based tokens, talked about when and why to use them, and covered a few "gotchas." We then revisited cookie handling between our front and backend parts before, finally, moving on to Vue Router.

Closing off the chapter with Vue Router, we looked at adding navigation guards, looked at how we can use meta values to enhance our development experience and mark pages for authorization, before finishing off by setting up our catch-all error-handling route so that our users have a great experience.

In the next chapters, we'll look at getting our app into production and getting ready for our first users.

Part 4:
Release and Deployment

The objective of this part of the book is to learn about the application release process and cloud deployments as part of the development process.

This part includes the following chapters:

11
Feature Flags

In this chapter, we will learn about feature flags, what they are, how to use them, and the benefits of using them. Using feature flags is not mandatory for applications. However, as application complexity increases, the need for feature flags will arise.

There are many different features provided by feature flags, but in this chapter, we will focus on how to use feature flags to enable/disable certain features in an application. We will be using an open source, simple version of the feature flag server to demonstrate the integration for both frontend and backend services.

In this chapter, we'll cover the following topics:

- Understanding what feature flags are all about
- Installing an open source feature flag server
- Enabling/disabling features using feature flags
- Integrating feature flags for frontend and backend services

Technical requirements

All the source code explained in this chapter can be found at `https://github.com/PacktPublishing/Full-Stack-Web-Development-with-Go/tree/main/chapter11`.

This chapter uses the cURL tool to perform HTTP operations. The tool is available for different operating systems and can be downloaded from `https://curl.se/download.html`.

An introduction to feature flags

In the current rapidly changing world, developers need to make changes and roll out new features almost every single day, if not quicker. Sometimes, this requires features to be built even before there are any user needs. Having the ability to deploy features into production without disruption is the holy grail of software development.

Features that are deployed to production may or may not be made available to users; this all depends on tactical decisions on the business side. Developers will keep on releasing features to production and, when the time is right, the feature will be made available with a click of a button from the business side. This kind of facility is provided by the feature flag.

In simple terms, feature flags are like on/off switches that we can use to enable/disable features in our applications without creating disruption. Enabling features will also allow companies to strategically enable or disable features depending on the market and users' needs, which can impact the bottom line of a company.

As a tool, feature flags not only provide the ability to run on/off features but there are also many other benefits you can also get out of it:

- Testing features for certain demographics based on certain conditions such as geographical location, user's age, and so on
- Splitting of the traffic request based on network condition
- Conducting UX experiments to understand which design works well

In this chapter, we will look at an open source project feature flag tool to demonstrate how to use and integrate it.

Feature flag configuration

You can use feature flags by deploying them in your infrastructure or by using software-as-a-service solutions such as LaunchDarkly, Flagsmith, and many other available solutions. Each of the solutions provides its own API, which needs to be integrated into your application. This means that your application is tied to the solution that you choose. There is no one-size-fits-all solution; it all depends on what kind of features you need for your application.

Let's take a look at different kinds of configuration for using feature flags. *Figure 11.1* shows the simplest way to use feature flags.

Figure 11.1: A web client using feature flags

The web client will enable or disable the user interface depending on the feature flag. For example, in an application, a particular menu selection can be enabled when the feature flag related to the menu is turned on.

Figure 11.2 shows a different configuration where the web client will call different microservices, depending on which feature flag is turned on/off:

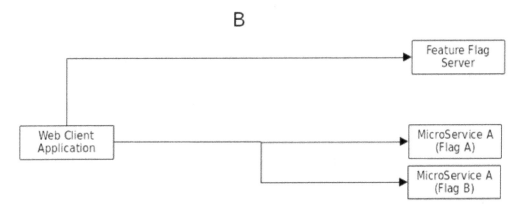

Figure 11.2: Feature flag microservices

In the preceding example, the web client calls microservice A when feature flag A is turned on.

Another interesting configuration is shown in *Figure 11.3*, where the main microservice will determine which user interface will be returned back to the web client, depending on which feature flag has been configured:

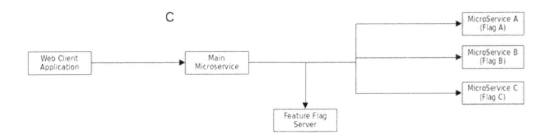

Figure 11.3: The feature flags for microservices

In the above example, the web client will get a different response to render if the main microservice detects that feature flag C has been enabled, which will get the response from MicroService C.

So, as we can see, there are different ways to use feature flags and different places to use them. It all comes down to what will be needed for your application.

In the next section, we will look at using an open source feature flag server to enable/disable buttons in a sample web app.

Use cases for using feature flags

Feature flags are not limited to flags that can be configured to turn on/off features inside applications; there are many more features and capabilities. In this section, we will look at the features provided in a full-blown feature flag server:

- **Segment targeting** – Imagine you are building a feature that you want to test on a group of users in your application. For example, you may want a certain group of users that are based in the USA to use the checkout feature based on PayPal.

- **Risk mitigation** – Building product features does not guarantee that a feature will bring in more users. New features can be released and, with time and more analysis, if it is found that the feature is providing a bad user experience, it will be turned off as part of the risk mitigation process.

- **Gathering feedback before launch** – Using a targeted rollout for a certain group of users, it is possible to get feedback as early as possible from the user. The feedback will provide insight for the team to decide whether the feature indeed provides any additional benefit to the user or not.

Now we have a good understanding of the use cases for feature flag, we will look at installing the feature flag server in the next section.

Installing the feature flag server

We are going to use an open source feature flag server. Clone the project from the `github.com/nanikjava/feature-flags` repository as follows:

```
git clone https://github.com/nanikjava/feature-flags
```

From your terminal, change the directory into the project directory and build the server using the following command:

```
go build -o fflag .
```

We are using the `-o` flag to compile the application and output it to a filename called `fflag`. Now that the server has been compiled and is ready to use, open a separate terminal window and run the server as follows:

```
./fflag
```

You will see the following output:

```
2022/07/30 15:10:38 Feature flag is up listening on  :8080
```

The server is now listening on port 8080. Now, we need to add a new feature flag for our web app, and the key is called `disable_get`. The way to do this is to use the `curl` command line to send data using POST as follows:

```
curl -v -X POST http://localhost:8080/features -H "Content-
Type:
application/json" -d '{"key":"disable_get","enabled":false,
"users":[],"groups":["dev","admin"],"percentage":0}'
```

The `curl` command is calling the `/features` endpoint and passing the JSON data. Once this has completed successfully, you will see the following output:

```
{
    "key": "disable_get",
    "enabled": false,
    "users": [],
    "groups": [
      "dev",
      "admin"
    ],
    "percentage": 0
}
```

The JSON output shows that the feature flag server now has a new key called `disable_get`, and it is disabled, as indicated by the flag `enabled: false`. The output should look as follows, showing that the data has been successfully added:

```
*    Trying 127.0.0.1:8080...
* Connected to localhost (127.0.0.1) port 8080 (#0)
...
* Mark bundle as not supporting multiuse
< HTTP/1.1 201 Created
...
< Content-Length: 89
<
```

```
{"key":"disable_get","enabled":false,"users":[],"groups":
["dev","admin"],"percentage":0}
* Connection #0 to host localhost left intact
```

The feature flag server is ready with the data we need. In the next section, we will look at using the flag inside our web app.

The high-level architecture of feature flags

Figure 11.4 shows the architecture of the open source feature flag server at a high level.

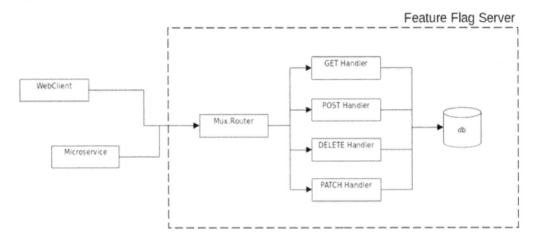

Figure 11.4: The high-level architecture

Looking at the diagram, the server uses mux.Router to route for different HTTP requests such as GET, POST, DELETE, and PATCH. The server uses an internal database as persistent storage for the feature flags that the application requires.

The server is accessible via HTTP request calls that can be made from both web clients and microservices using the normal HTTP protocol.

Integration of the feature flag

After we have installed the feature flag server, we want to start using it in our application. In this section, we will look at integrating the feature flag to enable/disable certain user interface elements in the frontend and to call only the backend services from our server that are enabled.

Web application

The sample app we are going to use can be found inside the `chapter11/frontend-enable-disable` folder; the sample app demonstrates how to use the feature flag to enable/disable the user interface button. Open the terminal and change into the `chapter11/frontend-enable-disable` directory to run the web app as follows:

```
npm install
npm run dev
```

The commands will install all the necessary packages and then run the web app. Once the command completes, open your browser and type `http://localhost:3000` in the address bar. You will see the web app shown in *Figure 11.5*.

Figure 11.5: The initial view of the web app using the feature flag

What you are seeing in *Figure 11.5* is that one of the buttons has been disabled. This is based on the flag value that we set in the previous section. The flag data looks as follows:

```
{
  "key": "disable_get",
  "enabled": false,
  "users": [],
  "groups": [
    "dev",
    "admin"
  ],
  "percentage": 0
}
```

The `disable_get` key is the flag key we added to the server and the `enabled` field is set to `false`, which means that the button is disabled. Let's change the `enabled` field to `true` and let's see how the web page changes.

Use the following command in a terminal to update the data:

```
curl -v -X PATCH http://localhost:8080/features/disable_get
-H "Content-Type: application/json" -d '{"key":"disable_
get","enabled":true}'
```

The `curl` command updates the `enabled` field to `true`. Refresh the browser page and you will see the button is enabled, as shown in *Figure 11.6*.

Figure 11.6: The Click to Get button is enabled

The following code snippet from the `HelloWorld.vue` file takes care of reading the key from the server, using it to enable/disable the button:

```
. . .

<script>
import axios from 'axios';

export default {
  data() {
    return {
      enabled: true
    }
  },
  mounted() {
    axios({method: "GET", "url":
      "http://localhost:8080/features/disable_get"}).then(result
        => {
      this.enabled = result.data.enabled
      console.log(result);
    }, error => {
```

```
      console.error(error);
   });
 }
}
</script>

<template>
 <div  v-if="enabled" class="flex space-2 justify-center">
   ...
   </button>
 </div>
 ...
```

In the next section, we will look at using the feature flag to enable/disable certain features on the backend service.

Microservice integration

In this section, we will use the feature flag to enable/disable certain services. This will give the application the flexibility to access only the services that are currently enabled. *Figure 11.7* shows how the microservice will be structured. The application is inside the chapter11/multiple-service folder.

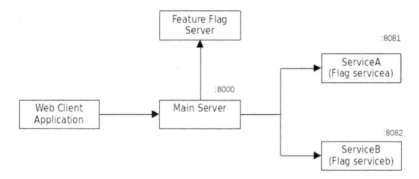

Figure 11.7: The microservice structure for the feature flag

Following the steps from the previous section to run the feature flag server, use the following command to create the flags:

```
curl -v -X POST http://localhost:8080/features -H "Content-
Type: application/json" -d '{"key":"serviceb", "enabled":true,
```

```
"users":[],"groups":["dev","admin"],"percentage":0}'

curl -v -X POST http://localhost:8080/features -H "Content-
Type: application/json" -d '{"key":"servicea", "enabled":false,
"users":[],"groups":["dev","admin"],"percentage":0}'
```

The command creates two keys: `servicea` and `serviceb`. For now, `servicea` is disabled, while `serviceb` is enabled. Once the feature flags have been set up, we are going to run different services:

- **Main server** – open the terminal and make sure you are inside the `chapter11/multiple-service/mainserver` directory. Run the main server with the following command:

  ```
  go run main.go
  ```

  The main server will run on port 8080.

- `servicea` – open the terminal and change the directory to `chapter11/multiple-service/servicea`. Run the service with the following command:

  ```
  go run main.go
  ```

  `servicea` will run on port 8081.

- `serviceb` – open the terminal and change the directory to `chapter11/multiple-service/serviceb`. Run the service with the following command:

  ```
  go run main.go
  ```

  `serviceb` will run on port 8082.

We now have three different services running on ports 8080, 8081, and 8082. Open your browser and access the service using `http://localhost:8000`. You will get a response like the following:

```
{"Message":"-ServiceB active"}
```

The response sent back came from `serviceb` as `servicea` is disabled, as per the configuration from the feature flag server. Now, let's turn on the flag for `servicea` using the following command:

```
curl -v -X PATCH http://localhost:8080/features/servicea -H
"Content-Type: application/json" -d '{"enabled":true}'
```

Restart the main server by force-stopping it using *Ctrl* + *C*. Re-run it using the same command discussed previously. Open your browser and access the service using `http://localhost:8000`. You should get a result like the following:

```
{"Message":"ServiceA active-ServiceB active"}
```

We get responses from both services now that both have been enabled.

Let's take a look at the code to understand how the feature flag is used. The following snippet shows part of the code to start the server:

```go
. . .
func main() {
  port := ":8000"
    . . .
  wg := &sync.WaitGroup{}

  wg.Add(1)
  go func(w *sync.WaitGroup) {
      defer w.Done()
      serviceA = checkFlags("servicea")
      serviceB = checkFlags("serviceb")
  }(wg)
  wg.Wait()

  http.ListenAndServe(port, rtr)
}
```

The code calls the feature flag server to get flag information for `servicea` and `serviceb` in a goroutine. Once completed, it starts the server to listen on port `8000`. The state of the services is stored inside the `servicea` and `serviceb` variables, which will be used in other parts of the code, as shown in the following code snippet:

```go
func handler() http.HandlerFunc {
  type ResponseBody struct {
      Message string
  }
  return func(rw http.ResponseWriter, req *http.Request) {
      var a, b string
      if serviceA {
          a = callService("8081")
      }
      if serviceB {
          b = callService("8082")
      }
```

```
    json.NewEncoder(rw).Encode(ResponseBody{
        Message: a + "-" + b,
    })
  }
}
```

The `handler()` method is called when you access the server on port 8000. Inside the code, as can be seen, it calls the service only when it is enabled. Once the service is called, the results from the service are combined and sent back to the client as a single JSON response.

The following code snippet shows how to access the feature flag server to extract the different flags. It uses a normal HTTP GET call:

```
func checkFlags(key string) bool {
    ...
    requestURL := fmt.Sprintf("http://localhost:%d/features/%s",
        8080, key)
    res, err := http.Get(requestURL)
    ...

    resBody, err := ioutil.ReadAll(res.Body)
    if err != nil {
        log.Printf("client: could not read response body: %s\n",
            err)
        os.Exit(1)
    }

    ...

    return f.Enabled
}
```

The code is calling the feature flag server by getting each key that we are interested in. So, in the case of the sample, we are calling using the following URLs:

```
http://localhost:8080/features/servicea
http://localhost:8080/features/serviceb
```

For example, when calling `http://localhost:8080/features/servicea`, the code will get the following JSON response from the feature flag server:

```
{
    "key": "servicea",
    "enabled": true,
    "users": [],
    "groups": [
        "dev",
        "admin"
    ],
    "percentage": 0
}
```

The `checkFlags()` function is interested only in the `enabled` field, which will be unmarshalled into the `FeatureFlagServerResponse` struct as shown below:

```
func checkFlags(key string) bool {
    type FeatureFlagServerResponse struct {
        Enabled bool `json:"enabled"`
    }
    ...
    var f FeatureFlagServerResponse
    err = json.Unmarshal(resBody, &f)
    ...
}
```

After successfully converting the JSON to a struct, it will return the `Enabled` value as shown here:

```
func checkFlags(key string) bool {
    ...

    return f.Enabled
}
```

We have come to the end of the chapter. In this section, we looked at integrating the feature flag in different scenarios such as inside web applications as well as using it as a feature toggle for accessing different microservices. There are other use cases where feature flags can be used, such as enabling/disabling performance metrics in production and enabling tracing in production for troubleshooting bugs.

Summary

In this chapter, we learned about feature flags, including what they are used for and how to use them. We learned how to install a simple feature flag server and saw how to integrate it with our sample application.

We went through the steps of using feature flags in two different use cases – integrating it by checking on the flag to enable/disable a button in our frontend and in the backend to call different microservices. Using feature flags to enable or disable certain services gives the application flexibility on what response will be sent back to the frontend, which gives developers the ability to allow access to certain services as and when needed.

In the next chapter, we will look at building continuous integration by exploring the different features provided by GitHub.

12

Building Continuous Integration

Building web applications to solve a problem is great, but we also need to make the applications available to users so they can start using them. As developers, we write code. But, at the same time, this code will need to be built or compiled so that it can be deployed, allowing users to use it. We need to understand how we can build our web application automatically, without requiring any manual process to work through. This is what we are going to talk about in this chapter. We will look at what is known as **continuous integration** (**CI**).

CI is a practice or process for automating the integration of code from different contributors into a project. CI allows developers to frequently merge code into a code repository where it will be tested and built automatically.

In this chapter, we will learn about the following for CI:

- GitHub workflows
- Using GitHub Actions
- Publishing to GitHub Packages

Technical requirements

The source code for this chapter can be found at `https://github.com/PacktPublishing/Full-Stack-Web-Development-with-Go/tree/main/chapter12`. In this chapter, we will also be using another repository when setting up CI for explanatory purposes. The repository is `https://GitHub.com/nanikjava/golangci`.

Importance of CI

You can think of CI as one aspect of your development process. The main reason why this is important is to allow you, as developers, to ensure that all code that is committed into a central code repository is tested and validated.

This becomes crucial when you are working in a team environment where multiple developers are working on the same project. Having proper CI will give developers peace of mind and assurance that all code they are using can be compiled properly and that automated test cases have been run successfully. Imagine that you have to check out some projects from GitHub, but when you try to compile and run some test cases, it fails; it would be a disaster as you would have to spend time fixing things, but if the project had a proper CI process set up, it would ensure all the committed code would compile correctly and test cases would pass.

Even when working as a solo developer on a project, it is highly recommended to have CI in place. The minimum benefit you will get from this is the assurance that your code can be built correctly. This also makes sure that any local dependencies related to your local machine that have been accidentally added to the code are detected when a build failure occurs.

In the next section, we will look at building our CI using GitHub by going through the different steps required to have CI for our web application.

Setting up GitHub

In this section, we will explain the different things that need to be prepared to get automated CI in GitHub. To gain a better understanding of the CI process, it is recommended that you create your own separate GitHub repository and copy everything inside the chapter12 directory to the new repository. Initially, when the nanikjava/golangci repository is created, it will look similar to *Figure 12.1*.

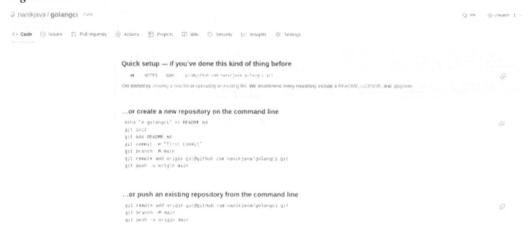

Figure 12.1: A fresh GitHub repo

For this chapter, we have set up a separate repository (`https://GitHub.com/nanikjava/` `golangci`) that we will use as a reference guide for the discussions in this chapter. We will go through the steps of creating a simple GitHub workflow in the repository. A GitHub workflow is a set of instructions that run one or more jobs. The instructions are defined in a YAML file with the extension of `.yaml` in the `.GitHub/workflows` directory of the repository.

You can define multiple workflows for your repository that perform different automated processes. For example, you can have one workflow file to build and test your application and another for deploying the application to a central location.

Let's create a simple workflow file inside the new repository by following the steps below:

1. From your repository, click on the **Actions** menu. This will bring you to the **Get Started with GitHub Actions** page, as shown in *Figure 12.2*.

Figure 12.2: The Get started with GitHub Actions page

2. Click on the **set up a workflow yourself** link. This will take to you a new page where you can start writing your workflow, as shown in *Figure 12.3*.

Figure 12.3: The create a new workflow screen

For now, we are going to create a simple workflow that we can use from GitHub. The workflow can be found at `https://docs.GitHub.com/en/actions/quickstart`. Copy and paste the workflow, as shown in *Figure 12.4*.

Figure 12.4: A sample GitHub workflow .yaml file

3. Commit the file by clicking on the **Start commit** button, as shown in *Figure 12.5*. After filling in all the commit information, click on the **Commit new file** button.

Figure 12.5: The commit message for a .yaml file

Your repo now has a new GitHub workflow file. If you select the **Actions** menu again, this time you will see that your screen looks like *Figure 12.6*. The screen shows that GitHub has run the workflow successfully.

Figure 12.6: GitHub has successfully run the workflow

We can look at the workflow results by clicking on the **Create main.yaml** link. You will see that the output indicates that the **Explore-GitHub-Actions** job was successfully run, as shown in *Figure 12.7*.

Figure 12.7: The Explore-GitHub-Actions step has been successfully run

After clicking on the **Explore-GitHub-Actions** jobs link, the output will be as shown in *Figure 12.8*.

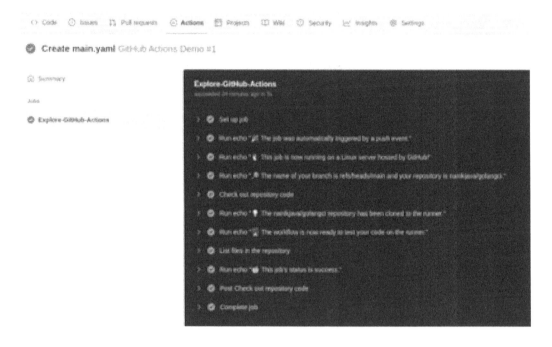

Figure 12.8: The Explore-GitHub-Actions job output

The workflow that we created in this section is actually the GitHub Actions workflow. We will look at this in more depth in the next section.

GitHub Actions

What is GitHub Actions? It is a platform that allows you to automate the complete integration and delivery of your project by automating the build, test, and deployment processes. GitHub Actions also gives you the ability to automate workflow processes such as pull requests, issue creation, and others.

We have now successfully created our first GitHub workflow. Let's take a look at the workflow file to get an understanding of which GitHub Actions we are using. The workflow file we will use is as follows:

```
name: GitHub Actions Demo
on: [push]
jobs:
  Explore-GitHub-Actions:
    runs-on: ubuntu-latest
    steps:
```

```
- run: echo "🎉 The job was automatically triggered by a
           ${{ GitHub.event_name }} event."
- run: echo "🐧 This job is now running on a ${{ runner.
           os }}
              server hosted by GitHub!"
- run: echo "🔎 The name of your branch is ${{ GitHub.
           ref }} and your repository is ${{ GitHub.
           repository }}."
- name: Check out repository code
  uses: actions/checkout@v3
- run: echo "💡 The ${{ GitHub.repository }} repository
           has been cloned to the runner."
- run: echo "🖥 The workflow is now ready to test your
           code on the runner."
- name: List files in the repository
  run: |
     ls ${{ GitHub.workspace }}
- run: echo "🍏 This job's status is ${{ job.status }}."
```

The following table explains the different configurations in the file:

Configuration key	Explanation
Name	The generic name we give to the workflow that will be used as a label for viewing the results on the Actions page.
On	Indicates to GitHub what kind of Git operation will trigger the workflow. In the example, it's push. This means that the workflow will be triggered every time the Git push operation is detected in the repository. Different Git event operations can be seen in the GitHub docs: https://docs.GitHub.com/en/actions/using-workflows/triggering-a-workflow#using-events-to-trigger-workflows.
Jobs	The workflow is made up of one or more jobs. These jobs are run in parallel by default. Jobs can be thought of as a single task that you want to do on your code. In our example, we named the job Explore-GitHub-Actions and it performs tasks defined by the run configuration.

Configuration key	Explanation
runs-on	Defines the runner that we want to use. The runner is the machine that you choose to run your workflow on. In our example, we are using the ubuntu-latest machine, or, in other words, we want to use a machine that runs the latest version of Ubuntu. A complete list of runners can be seen in the following link: `https://docs.GitHub.com/en/actions/using-jobs/choosing-the-runner-for-a-job`.
Steps	Each job contains a sequence of tasks called steps. A step is where you define the operation you want to perform for the workflow. In our example, we defined several steps such as `run` where we just print out information.

Now, we are going to take a look at the GitHub Action workflow we have for the sample application. The workflow can be found inside the `chapter12/.GitHub/workflows/build.yml` file, as shown here:

```yaml
name: Build and Package
on:
 push:
   branches:
     - main
 pull_request:

jobs:
 lint:
   name: Lint
   runs-on: ubuntu-latest
   steps:
     - name: Set up Go
       uses: actions/setup-go@v1
       with:
         go-version: 1.18

     - name: Check out code
       uses: actions/checkout@v1

     - name: Lint Go Code
       run:
```

```
        curl -sSfL
        https://raw.GitHubusercontent.com/golangci/golangci-
            lint/
        master/install.sh | sh -s -- -b $(go env GOPATH)/bin
        $(go env GOPATH)/bin/golangci-lint run

  build:
    name: Build
    runs-on: ubuntu-latest
    needs: [ lint ]
    steps:
      - name: Set up Go
        uses: actions/setup-go@v1
        with:
          go-version: 1.18

      - name: Check out code
        uses: actions/checkout@v1

      - name: Build
        run: make build
```

We will go now through this line by line to understand what the workflow is doing. The following snippet tells GitHub that the workflow will be triggered when source code is pushed to the main branch:

```
name: Build and Package
on:
  push:
    branches:
      - main
```

The next snippet shows the different jobs that GitHub will run when the event is detected; in this case, the lint and build jobs. The job will be run on an Ubuntu machine, as specified by the runs-on configuration:

```
jobs:
  lint:
    name: Lint
```

```
    runs-on: ubuntu-latest
    steps:
        ...


  build:
    name: Build
    runs-on: ubuntu-latest
    needs: [ lint ]
    steps:
        ...
```

The defined jobs are made up of the steps shown in the following snippet:

```
    ...


jobs:
  lint:
    ...
    steps:
        - name: Set up Go
          uses: actions/setup-go@v1
          with:
            go-version: 1.18


        - name: Check out code
          uses: actions/checkout@v1


        - name: Lint Go Code
          run: |
            curl -sSfL
            https://raw.GitHubusercontent.com/golangci/golangci-
                lint/
            master/install.sh | sh -s -- -b $(go env GOPATH)/bin
            $(go env GOPATH)/bin/golangci-lint run


  build:
    ...
```

```
steps:
  - name: Set up Go
    uses: actions/setup-go@v1
    with:
      go-version: 1.18

  - name: Check out code
    uses: actions/checkout@v1

  - name: Build
    run: make build
```

The explanation of the steps performed for the `lint` job is as follows:

1. Set up a Go 1.18 environment using the `actions/setup-go` GitHub Action.
2. Check out the source code using the `actions/checkout` GitHub Action.
3. Perform a linting operation on the source code. The shell script will install the `golangci-lint` tool and run it using the `golangci-lint run` command.

The other `build` job will perform the following steps:

1. Set up a Go 1.18 environment using the `actions/setup-go` GitHub Action.
2. Check out the source code using the `actions/checkout` GitHub Action.
3. Build the application by executing the `make build` command.

Each step defined inside a job uses GitHub Actions that perform different operations such as checking out code, running shell script, and setting up the environment for compiling the Go application.

In the next section, we will look at GitHub Packages and how to use them to deploy the Docker image that we will build for our application.

Publishing Docker images

After developing your application, the next step is to deploy the application so that your user can start using it. To do this, you need to package your application. This is where Docker comes into the picture. Docker is a tool that is used to package your application into a single file, making it easy to deploy into a cloud environment such as Amazon, Google, and so on. We will look at Docker images and containers in depth in *Chapter 13, Dockerizing an Application*. We will look at the file with which we configure Docker, called the `Dockerfile`. We will briefly look at what this file does.

Dockerfile

`Dockerfile` is the default filename used to name a file that contains instructions for building an image for your application. The `Dockerfile` contains instructions on steps for Docker to perform to package your application into a Docker image.

Let's take a look at the `Dockerfile` that we have inside the `Chapter12` directory:

```
# 1. Compile the app.
FROM golang:1.18   as builder
WORKDIR /app
COPY . .
RUN CGO_ENABLED=0 GOOS=linux go build -a -o bin/embed

# 2. Create final environment for the compiled binary.
FROM alpine:latest
RUN apk --update upgrade && apk --no-cache add curl
ca-certificates && rm -rf /var/cache/apk/*
RUN mkdir -p /app

# 3. Copy the binary from step 1 and set it as the default
command.
COPY --from=builder /app/bin/embed /app
WORKDIR /app
CMD /app/embed
```

There are three major steps to package the application:

1. Compile our Go application into a binary file called `embed`.

2. Create an environment that will be used to run our application. In our example, we are using an environment or operating system called `alpine`.

3. Copy the binary that was built in the first step into the new environment that we set up in the second step.

We will use the `Dockerfile` in the next section to store the image in GitHub Packages.

GitHub Packages

GitHub Packages is a service provided by GitHub that allows developers to host their packages. These packages can be accessed either by your team or made available to the general public. We will use this service to publish our Docker image and make it available to be consumed by the public.

There are a few things we need to set up before we can deploy our Docker image into GitHub Packages. This section will walk you through the steps required to set up your repository. We will use `GitHub.com/nanikjava/golangci` as a reference in this section.

You can access GitHub Packages from your repository by clicking on the **Packages** link, as shown in *Figure 12.9*.

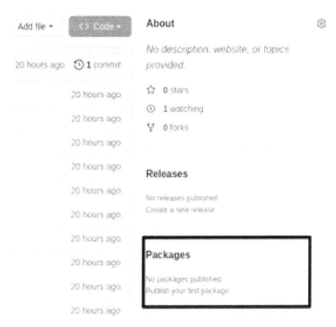

Figure 12.9: Access to GitHub Packages

Once you click on the **Packages** link, you will be shown a screen similar to that in *Figure 12.10*. There will be no **Packages** displayed as we have not yet published any.

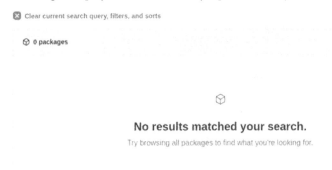

Figure 12.10: The GitHub Packages page

In the next section, we will look at how to publish the Docker images that we turn into packages on GitHub Packages.

Publishing to GitHub Packages

Security is an important part of GitHub. In order to be able to write Docker images into GitHub Packages, let's try to understand what is required. Every time GitHub runs a workflow, a temporary token is assigned to the workflow that can be used as an authentication key, allowing GitHub Actions to perform certain operations. This key is known as `GITHUB_TOKEN` internally.

The `GITHUB_TOKEN` key has default permissions that can be made restrictive, depending on your organization's needs. To see the default permissions, click on the **Settings** tab from your repository, as shown in *Figure 12.11*.

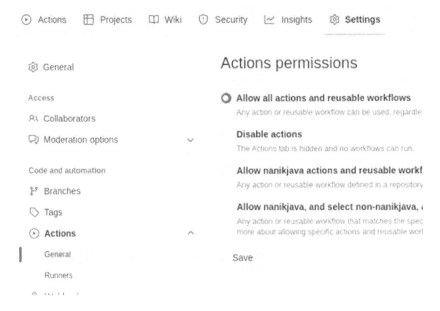

Figure 12.11: The Actions menu from Settings

Click on the **Actions** menu and select **General**. You will be shown the default permissions, as shown in *Figure 12.12*. As you can see, the default permissions are **Read and write** for the workflow.

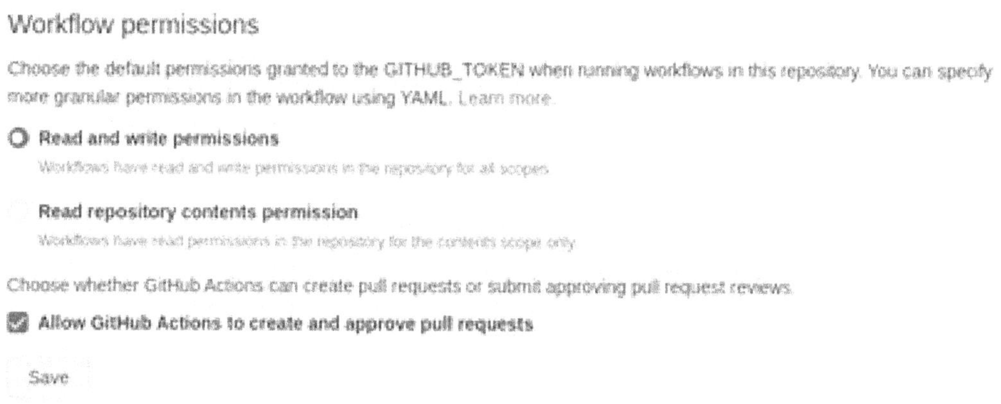

Workflow permissions

Choose the default permissions granted to the GITHUB_TOKEN when running workflows in this repository. You can specify more granular permissions in the workflow using YAML. Learn more.

⭕ **Read and write permissions**
  Workflows have read and write permissions in the repository for all scopes.

Read repository contents permission
  Workflows have read permissions in the repository for the contents scope only.

Choose whether GitHub Actions can create pull requests or submit approving pull request reviews.

☑ **Allow GitHub Actions to create and approve pull requests**

Save

Figure 12.12: The GITHUB_TOKEN default permissions

The workflow that we going to look at can be found inside `chapter12/.GitHub/workflows/builddocker.yml` and looks like the following:

```yaml
name: Build Docker Image
on:
 push:
   branches:
      - main
 pull_request:

env:
 REGISTRY: ghcr.io
 IMAGE_NAME: ${{ GitHub.repository }}

jobs:
 push_to_GitHub_registry:
   name: Push Docker image to Docker Hub
   runs-on: ubuntu-latest
   steps:
    ...

    - name: Log in to the Container registry
      uses: docker/login-action@v2
      with:
        registry: ${{ env.REGISTRY }}
```

```
      username: ${{ GitHub.actor }}
      password: ${{ secrets.GITHUB_TOKEN }}


  - name: Build and push Docker image
    uses: docker/build-push-action@v3
    with:
      context: .
      file: ./Dockerfile
      push: true
      tags: ${{ env.REGISTRY }}/${{ env.IMAGE_NAME
            }}/chapter12:latest
```

The workflow performs the following steps in order to publish the Docker image:

1. The workflow logs in to the registry (GitHub Packages) using the `docker/login-action@v2` GitHub Action. The parameters supplied to the GitHub Action are `username`, `password`, and `registry`.

2. The `username` is the GitHub username, which triggers the workflow process. The `registry` parameter will be value from the `REGISTRY` environment variable, which will be - `ghcr.io`. The `password` field will be automatically populated using `secrets.GITHUB_TOKEN`.

3. The last step is to build and publish the Docker image using the `docker/build-push-action@v3` GitHub Action. The parameter passed to the GitHub Action is the *file* that will be used to build the Docker image. In our case, it's called `Dockerfile`. The tag name used to tag or label the Docker image will look like `ghcr.io/golangci/chapter12:latest`.

Now that we have everything set up, the next time you push any code changes into the `main` branch, the workflow will run. An example of a successful run can be seen in *Figure 12.13*.

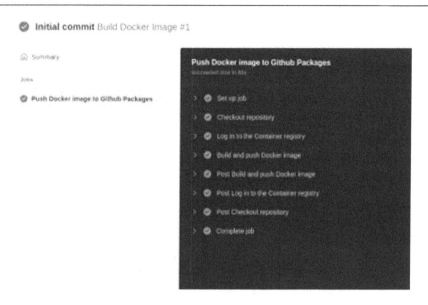

Figure 12.13: A successful workflow run publishing a Docker image

The Docker image can be seen on the GitHub Packages page, as shown in *Figure 12.14*.

Figure 12.14: The chapter12 Docker image inside GitHub Packages

In the next section, we will look at downloading our newly created Docker image and using it locally.

Pulling from GitHub Packages

We have successfully set up CI for our application. Now, we have to test whether the Docker image that was run as part of the CI process has successfully built our application's Docker image.

Our Docker image is hosted inside GitHub Packages, which is made public by default as our repository is a public repository. *Figure 12.14* shows the Docker images that are available to be used, including the command to pull the image locally. Open your terminal, then run the following command:

```
docker pull ghcr.io/nanikjava/golangci/chapter12:latest
```

You will get the following output:

```
latest: Pulling from nanikjava/golangci/chapter12
213ec9aee27d: Already exists
3a904afc80b3: Pull complete
561cc7c7d83b: Pull complete
aee36b390937: Pull complete
4f4fb700ef54: Pull complete
Digest: sha256:a355f55c33a400290776faf20b33d45096eb19a6431fb0b3
f723c17236e8b03e
Status: Downloaded newer image for ghcr.io/nanikjava/golangci/
chapter12:latest
```

The image has been downloaded to your local machine. Run the Docker image using the following command:

```
docker run -p 3333:3333 ghcr.io/nanikjava/golangci/chapter12
```

You know that the container is running when you see the following output:

```
2022/08/18 08:03:10 Server Version : 0.0.1
```

Open your browser and enter http://localhost:3333 into the browser address bar. You will see the login page. We have successfully completed our CI process and are able to run the Docker image that we have built.

Summary

In this chapter, we explored CI, developed an understanding of why it is important, and the benefits we get by setting up an automated CI process for a project. We learned to set up a GitHub repository to prepare our CI process and also learned to write a GitHub Actions workflow that enables us to automate a number of steps for our application.

Using GitHub Actions, we were able to build our application into an executable binary. This is performed every time we push code into the repository. We learned about building Docker images for our application and the benefits we get by packaging our application as a Docker image.

We learned about GitHub Packages and how to configure it to allow us to push our Docker images to a central location. Having our application packaged as a Docker image makes it easy for us to test our application anywhere. We don't have to worry about building the source code as everything is packaged together into a single Docker image file.

In the next chapter we will learn on how to package our application as container, which will make it easy to deploy as a single image and allow us to deploy application in the cloud using different cloud providers.

13

Dockerizing an Application

In this chapter, we will learn about Docker and how to package applications as Docker images. Understanding how to package your application as a Docker image will allow you to deploy the application in any kind of environment and infrastructure without having to worry about setting up the infrastructure to build your application. Building a Docker image will allow you to run your application anywhere you like: build once and deploy anywhere.

In this chapter, we will learn about the following key topics:

- Building a Docker image
- Running a Docker image
- Creating a Docker image from scratch
- Understanding the Docker image filesystem
- Looking at Docker Compose

Technical requirements

All the source code explained in this chapter can be checked out at `https://github.com/PacktPublishing/Full-Stack-Web-Development-with-Go/tree/main/chapter13`.

Installing Docker

Docker is an open source platform that is used for software development, making it easy to package and ship programs. Docker enables you to package your application and run it in different kinds of infrastructure such as cloud environments.

In this section, we will look at installing Docker on a local machine. Different operating systems have different steps for installing it. Refer to the Docker documentation for an in-depth installation guide relevant to your operating system, which can be found at `https://docs.docker.com/engine/install/`.

> **Note**
>
> This chapter was written on a Linux machine, so most of the command-line applications that
> are outlined are only available for Linux.

After taking the steps to install Docker on our development machine, the following are some of the
things we do to ensure that everything is working fine.

Use the following commands to check that the Docker engine is running:

```
systemctl list-units --type=service --state=running    | grep -i
docker && systemctl list-units --type=service --state=active    |
grep -i containerd
```

You will see the following output if the engine has been installed correctly:

```
   docker.service                          loaded     active running
Docker Application Container Engine
   containerd.service                 loaded     active running
containerd container runtime
```

The output shows two different services running – `docker.service` and `containerd.service`.
The `containerd.service` service takes care of launching the Docker image into a container
and ensuring that all the local machine services are set up to allow the container to run while the
`docker.service` service takes care of the management of the image and communication with
the Docker command-line tools.

Now that we know both services are running, let's use the command-line tools to check the communication
with the engine. Use the following command to communicate with the engine to list all the locally
available images – note you may need to have root privileges to do this so prefixing with `sudo` might
be required:

```
docker images
```

In our case, we get the output as shown in *Figure 13.1*, showing we have downloaded two images. In
your case, if this is your first time installing Docker, it will be empty.

```
         REPOSITORY    TAG        IMAGE ID         CREATED         SIZE
         redis         latest     bba24acba395     4 weeks ago     113MB
         postgres      latest     1ee973e26c65     4 weeks ago     376MB
```

Figure 13.1: Docker images on a local machine

We have successfully completed the Docker installation on the local machine. In the next section, we
will go into more detail about using Docker and understanding Docker images.

Using Docker

In this section, we will look at how to use Docker for day-to-day operations. Let's understand the concepts that are talked about when using Docker – images and the container:

- **Docker image**: This image is a file that contains our application, including all the relevant operating system files.

- **Container**: The image file is read and executed by the Docker engine. Once it runs on the local machine it is called a container. You can interact with the container using the Docker command-line tools.

We will look at using Docker to download and run a simple *Hello World* application using the following command:

```
docker run hello-world
```

Open your terminal and run the following command:

```
$ docker run hello-world
```

This command will download the image file (if none exists locally) and execute it. You will then see the following message:

```
Unable to find image 'hello-world:latest' locally
latest: Pulling from library/hello-world
2db29710123e: Pull complete
Digest: sha256:10d7d58d5ebd2a652f4d93fdd86da8f265f5318c6a73cc5b
6a9798ff6d2b2e67
Status: Downloaded newer image for hello-world:latest
```

Once the image has been downloaded and run as a container, it will print out the following output:

```
Hello from Docker!
This message shows that your installation appears to be working
correctly.

To generate this message, Docker took the following steps:
 1. The Docker client contacted the Docker daemon.
 ...
 ...

https://docs.docker.com/get-started/
```

Now that we have had a taste of how to run an image file as a container, we will explore Docker images more in the next section.

Docker images

Docker image files look like any other file on your local machine, except they are stored in a special format that can only be understood by Docker. Locally the image files are stored inside the `/var/lib/docker/image/overlay2` directory. To see what images are available, you can take a look at the `repositories.json` file, which looks as follows:

```
{
    "Repositories": {
        "hello-world": {
            "hello-world:latest":
                "sha256:feb5d9fea6a5e9606aa995e879d862b82
                5965ba48de054caab5ef356dc6b3412",
            "hello-world@sha256:
                10d7d58d5ebd2a652f4d93fdd86da8f265f5318c6a7
                3cc5b6a9798ff6d2b2e67":
                    "sha256:feb5d9fea6a5e9606aa995e879d862
                    b825965ba48de054caab5ef356dc6b3412"
        },
        "...
        "redis": {
            "redis:latest":
                "sha256:bba24acba395b778d9522a1adf5f0d6bba3e609
                4b2d298e71ab08828b880a01b",
            "redis@sha256:69a3ab2516b560690e37197b71bc61ba24
                            5aafe4525ebdec
            e1d8a0bc5669e3e2":
                "sha256:bba24acba395b778d9522a1adf5f0d6bba3
                e6094b2d298e71ab08828b880a01b"
        }
    }
}
```

Let's explore the Docker directories that host the image files further. We can get the image information using the following command:

```
docker images
```

The following output shows some information about the `hello-world` container:

```
REPOSITORY      TAG         IMAGE
ID              CREATED         SIZE
..
hello-world        latest      feb5d9fea6a5    7 months ago    13.3kB
..
```

The image ID for `hello-world` is `feb5d9fea6a5`. Let's try to find the image file inside `/var/lib/docker` using the following command:

```
sudo find /var/lib/docker -name 'feb5d9fea6a5*'
```

We will get the following output:

```
/var/lib/docker/image/overlay2/imagedb/content/sha256/feb5d9fea
6a5e9606aa995e879d862b825965ba48de054caab5ef356dc6b3412
```

Let's now look inside that file using the following command:

```
sudo cat /var/lib/docker/image/overlay2/imagedb/content/sha256/
feb5d9fea6a5e9606aa995e879d862b825965ba48de054caab5ef356dc6b3412
```

You will see the following output:

```
{
  "architecture": "amd64",
  "config": {
    ...
    ],
      ...
  },
    ...

    "Cmd": [
      "/bin/sh",
      "-c",
```

```
        "#(nop) ",
        "CMD [\"/hello\"]"
    ],
    "Image": "sha256:b9935d4e8431fb1a7f0989304ec8
                6b3329a99a25f5efdc7f09f3f8c41434ca6d",
    "Volumes": null,
    "WorkingDir": "",
    "Entrypoint": null,
    "OnBuild": null,
    "Labels": {}
},
"created": "2021-09-23T23:47:57.442225064Z",
"docker_version": "20.10.7",
"history": [
    {
        ...
    }
],
"os": "linux",
"rootfs": {
    "type": "layers",
    "diff_ids": [
        "sha256:e07ee1baac5fae6a26f30cabfe54a36d3402f96afda3
                18fe0a96cec4ca393359"
    ]
}
}
```

The following table outlines the meanings of some relevant fields from the preceding JSON output:

Field Name	Description
Cmd	This is the command that will be executed when the image file is run as a container. For the hello-world example, it will execute the hello executable when the container is launched.
rootfs	rootfs stands for *root filesystem*, which means it contains all the necessary operating system files that are required to start itself as a normal machine.

The JSON information we saw previously can also be viewed using the following command:

```
docker image inspect hello-world:latest
```

You will get output that looks as follows:

```
[
    {
        "Id": "sha256:feb5d9fea6a5e9606aa995e879d862b825
                965ba48de054caab5ef356dc6b3412",
        "RepoTags": [
            "hello-world:latest"
        ],
        "RepoDigests": [
            "hello-world@sha256:10d7d58d5ebd2a652
            f4d93fdd86da8f265f5318c6a73cc5b6a9798ff6d2b2e67"
        ],
        "Parent": "",
        "Comment": "",
        "Created": "2021-09-23T23:47:57.442225064Z",
        "Container": "8746661ca3c2f215da94e6d3f7dfdcafaff5
                    ec0b21c9aff6af3dc379a82fbc72",
        "ContainerConfig": {
            ...
            "Cmd": [
                "/bin/sh",
                "-c",
                "#(nop) ",
                "CMD [\"/hello\"]"
            ],
            "Image": "sha256:b9935d4e8431fb1a7f0989304ec86b
                    3329a99a25f5efdc7f09f3f8c41434ca6d",
            ...
        },
        ...
        "Architecture": "amd64",
        "Os": "linux",
```

```
            "Size": 13256,
            "VirtualSize": 13256,
            "GraphDriver": {
                "Data": {
                    "MergedDir":
                      "/var/lib/docker/overlay2/c0d9b295437ab
                      cdeb9caeec51dcbde1b11b0aeb3dd9e469f35
                      7889defed757d9/merged",
                    "UpperDir":
                      "/var/lib/docker/overlay2/c0d9b295437ab
                      cdeb9caeec51dcbde1b11b0aeb3dd9e469f357
                      889defed757d9/diff",
                    "WorkDir":
                      "/var/lib/docker/overlay2/c0d9b295437ab
                      cdeb9caeec51dcbde1b11b0aeb3dd9e469f357
                      889defed757d9/work"
                },
                "Name": "overlay2"
            },
                    ...]
```

One of the interesting pieces of information in the output is the `GraphDriver` field that points to the `/var/lib/docker/overlay2/c0d9b295437abcdeb9caeec51dcbde1b11b0aeb3dd9e469f357889defed757d9` directory containing the extracted Docker image. For hello-world, it will be the `hello` executable file, as shown next:

```
total 16
drwx--x---  3 root root 4096 Apr 30 18:36 ./
drwx--x--- 30 root root 4096 Apr 30 19:21 ../
-rw-------  1 root root    0 Apr 30 19:21 committed
drwxr-xr-x  2 root root 4096 Apr 30 18:36 diff/
-rw-r--r--  1 root root   26 Apr 30 18:36 link
```

Taking a look inside the `diff/` directory, we see the following executable file:

```
drwxr-xr-x 2 root root  4096 Apr 30 18:36 .
drwx--x--- 3 root root  4096 Apr 30 18:36 ..
-rwxrwxr-x 1 root root 13256 Sep 24  2021 hello
```

Now that we have a good understanding of how Docker images are stored locally, in the next section, we will look at using Docker to run the image locally as a container.

Running images as containers

In this section, we will look at running Docker images as containers and examine the different information that we can see when a container is running.

Start by running a database Docker image and look at what information we can get about the state of the container. Open the terminal window and run the following command to run Redis locally. Redis is an open source memory-based data store used to store data. Since data is stored in memory, it is fast compared to storing on disk. The command will run Redis, listening on port 7777:

```
docker run -p 7777:7777  -v /home/user/Downloads/redis-7.0-rc3/
data:/data redis --port 7777
```

Make sure you change the /home/user/Downloads/redis-7.0-rc3/data directory to your own local directory, as Docker will use this to store the Redis data file.

You will see the following message when the container is up and running:

```
1:C 05 May 2022 11:20:08.723 # oO0OoO0OoO0Oo Redis is starting
oO0OoO0OoO0Oo
1:C 05 May 2022 11:20:08.723 # Redis version=6.2.6, bits=64,
commit=00000000, modified=0, pid=1, just started
1:C 05 May 2022 11:20:08.723 # Configuration loaded
1:M 05 May 2022 11:20:08.724 * monotonic clock: POSIX clock_
gettime
1:M 05 May 2022 11:20:08.724 * Running mode=standalone,
port=7777.
...
1:M 05 May 2022 11:20:08.724 * Ready to accept connections
```

Let's use the Docker command-line tool to look at the running state of this container. In order to do that, we need to get the ID of the container by running the docker ps command; in our case, the output looks as follows:

```
CONTAINER ID    IMAGE       COMMAND               CREATED
    STATUS          PORTS
        NAMES
e1f58f395d06    redis       "docker-entrypoint.s…"    5 minutes
ago    Up 5 minutes    6379/tcp, 0.0.0.0:7777->7777/tcp, :::7777-
>7777/tcp    reverent_dhawan
```

The Redis container ID is e1f58f395d06. Using this information, we will use docker inspect to look at the different properties of the running container. Use docker inspect as follows:

```
docker inspect e1f58f395d06
```

You will get output that looks like the following:

```
[[
    {
        ...
        "Mounts": [
            {
                "Type": "bind",
                "Source": "/home/user/Downloads/redis-7.0-
                        rc3/data",
                "Destination": "/data",
                "Mode": "",
                "RW": true,
                "Propagation": "rprivate"
            }
        ],
        "Config": {
            ...
            "Env": [
                "PATH=/usr/local/sbin:/usr/local/bin:
                        /usr/sbin:/usr/bin:/sbin:/bin",
                "GOSU_VERSION=1.14",
                ...
            ],
            ...
        },
        "NetworkSettings": {
            ...
            "Ports": {
                "6379/tcp": null,
                "7777/tcp": [
                    {
                        "HostIp": "0.0.0.0",
```

```
                    "HostPort": "7777"
                },
                {
                    "HostIp": "::",
                    "HostPort": "7777"
                }
            ]
        },
        ...
        "Networks": {
            "bridge": {
                ...
            }
        }
    }
}
]
```

The output shows a lot of information about the running state of the Redis container. The main things that we are interested in are the network and the mount. The NetworkSettings section shows the network configuration of the container, indicating the network mapping parameter of the host to the container – the container is using port 7777, and the same port is exposed on the local machine.

The other interesting thing is the Mounts parameter, which points to the mapping of /home/user/ Downloads/redis-7.0-rc3/data to the /data local host directory inside the container. The mount is like a redirection from the container directory to the local machine directory. Using the mount ensures that all data is saved to the local machine when the container shuts down.

We have seen what a container is all about and how to look at the running state of the container. Now that we have a good understanding of images and containers, we will look at creating our own image in the next section.

Building and packaging images

In the previous section, we learned about Docker images and how to look at the state of a running container; we also looked at how Docker images are stored locally. In this section, we will look at how to create our own Docker image by writing a Dockerfile.

We will look at building the sample application inside the `chapter13/embed` folder. The sample application is the same one we discussed in *Chapter 4, Serving and Embedding HTML Content*. The application will run an HTTP server listening on port 3333 to serve an embedded HTML page.

The `Dockerfile` that we will use to build the Docker image looks as follows:

```
# 1. Compile the app.
FROM golang:1.18  as builder
WORKDIR /app
COPY . .
RUN CGO_ENABLED=0 GOOS=linux go build -a -o bin/embed


# 2. Create final environment for the compiled binary.
FROM alpine:latest
RUN apk --update upgrade && apk --no-cache add curl
ca-certificates && rm -rf /var/cache/apk/*
RUN mkdir -p /app


# 3. Copy the binary from step 1 and set it as the default
command.
COPY --from=builder /app/bin/embed /app
WORKDIR /app
CMD /app/embed
```

Let's step through the different parts of the command to understand what it is doing. The first step is to compile the application by using a pre-built Golang 1.18 Docker image. This image contains all the necessary tools to build a Go application. We specify `/app` as the working directory using the `WORKDIR` command, and in the last line we copy all the source files using the `COPY` command and compile the source code using the standard `go build` command line.

```
FROM golang:1.18  as builder
WORKDIR /app
COPY . .
RUN CGO_ENABLED=0 GOOS=linux go build -a -o bin/embed
```

After successfully compiling the application, the next step is to prepare the runtime environment that will host the application. In this case, we are using a pre-built Docker image of the Alpine Linux operating system. Alpine is a Linux distribution that is small in terms of size, which makes it ideal when creating Docker images for applications to run on.

The next thing we want to do is to make sure the operating system is up to date by using the - update upgrade command. This ensures that the operating system contains all the latest updates, including security updates. The last step is to create a new /app directory that will store the application binary:

```
FROM alpine:latest
RUN apk --update upgrade && apk --no-cache add curl
ca-certificates && rm -rf /var/cache/apk/*
RUN mkdir -p /app
```

The final step is to copy over the binary from the previous step, which we have labeled as builder, into the new /app directory. The CMD command specifies the command that will be run when the Docker image is executed as a container – in this case, we want to run our sample application embed specified by the parameter /app/embed:

```
COPY --from=builder /app/bin/embed /app
WORKDIR /app
CMD /app/embed
```

Now we have gone through what the Dockerfile is doing, let's create the Docker image. Use the following command to build the image:

```
docker build  --tag chapter13 .
```

You will see an output that looks like the following, showing the different steps and processes Docker is doing to build the image:

```
Sending build context to Docker daemon   29.7kB
Step 1/10 : FROM golang:1.18  as builder
 ---> 65b2f1fa535f
Step 2/10 : WORKDIR /app
 ---> Using cache
 ---> 7ab996f8148c
...
Step 5/10 : FROM alpine:latest
 ---> 0ac33e5f5afa
...
Step 8/10 : COPY --from=builder /app/bin/embed /app
...
Step 10/10 : CMD /app/embed
 ---> Using cache
```

```
 ---> ade99a01b92e
Successfully built ade99a01b92e
Successfully tagged chapter13:latest
```

Once you get the `Successfully tagged` message, the building process is complete, and the image is ready on your local machine.

The new image will be labeled `chapter13` and will look as follows when we use the `docker images` command:

```
REPOSITORY     TAG          IMAGE
ID        CREATED          SIZE
...
chapter13      latest          ade99a01b92e   33 minutes
ago       16.9MB
...
golang         1.18            65b2f1fa535f   14 hours
ago       964MB
...
hello-world    latest          feb5d9fea6a5   7 months
ago       13.3kB
```

Run the newly created image using the following command:

```
docker   run -p 3333:3333 chapter13
```

The command will run the image as a container, and using the -p port parameter, it exposes port 3333 inside the container to the same port 3333 on the host. Open your browser and type in http://localhost:3333 and you will see the HTML login page, as shown in *Figure 13.2*:

Figure 13.2: Web application served from a Docker container

In the next section, we'll understand about Docker Compose.

Docker Compose

Docker provides another tool called Docker Compose, allowing developers to run multiple containers simultaneously. Think about use cases where you are building a server that requires temporary memory storage to store cart information; this requires using an external application such as Redis, which provides an in-memory database.

In this kind of scenario, our application depends on Redis to function properly, which means that we need to run Redis at the same time we run our application. There are many other different kinds of use cases where there will be a need to use Docker Compose. The Docker Compose documentation provides a complete step-by-step guide on how to install it on your local machine: https://docs.docker.com/compose/install/.

Docker Compose is actually a file that outlines the different containers we want to use. Let's try to run the sample Docker Compose file that is inside the chapter13/embed folder. Open the terminal and make sure you are inside the chapter13/embed folder, then execute the following command:

```
docker compose -f compose.yaml up
```

You will get the following output:

```
[+] Running 7/7
 :: cache Pulled 11.6s
 :: 213ec9aee27d Already exists  0.0s
 :: c99be1b28c7f Pull complete    1.4s
 :: 8ff0bb7e55e3 Pull complete    1.8s
 :: 477c33011f3e Pull complete    4.8s
 :: 2bbc51a93257 Pull complete    4.8s
 :: 2d27eae19281 Pull complete    4.9s
[+] Building 7.3s (15/15) FINISHED
 => [internal] load build definition from Dockerfile    0.0s
 => => transferring dockerfile: 491B                    0.0s
 => [internal] load .dockerignore                       0.0s
 => => transferring context: 2B                         0.0s
 => [internal] load metadata for docker.io/library/
alpine:latest 0.0s
 => [internal] load metadata for docker.io/library/
golang:1.18    0.0s
 => [builder 1/4] FROM docker.io/library/
```

```
golang:1.18                 0.3s
 => [stage-1 1/5] FROM docker.io/library/
alpine:latest               0.1s
 => [internal] load build
context                              0.2s
 => => transferring context:
18.81kB                            0.0s
 => [stage-1 2/5] RUN apk --update upgrade && apk --no-cache
add curl ca-certificates && rm -rf /var/cache/apk/*    5.5s
 => [builder 2/4] WORKDIR /app  0.2s
 => [builder 3/4] COPY . .     0.1s
 => [builder 4/4] RUN CGO_ENABLED=0 GOOS=linux go build -a -o
bin/embed 6.4s
 => [stage-1 3/5] RUN mkdir -p /app  1.4s
 => [stage-1 4/5] COPY --from=builder /app/bin/embed /app   0.1s
 => [stage-1 5/5] WORKDIR /app  0.0s
 => exporting to image  0.1s
 => => exporting layers  0.1s
 => => writing image sha256:84621b13c179c03eed57a23c66974659eae
4b50c97e3f8af13de99db1adf4c06  0.0s
 => => naming to docker.io/library/embed-server  0.0s
[+] Running 3/3
 ⠿ Network embed_default     Created 0.1s
 ⠿ Container embed-cache-1    Created 0.1s
 ⠿ Container embed-server-1   Created 0.1s
Attaching to embed-cache-1, embed-server-1
embed-server-1 | 2022/09/10 06:24:30 Server Version : 0.0.1
embed-cache-1    | 1:C 10 Sep 2022 06:24:30.898 # oO0Oo00oO000o
Redis is starting oO0Oo00oO000o
embed-cache-1    | 1:C 10 Sep 2022 06:24:30.898 # Redis
version=7.0.4, bits=64, commit=00000000, modified=0, pid=1,
just started
...
embed-cache-1    | 1:M 10 Sep 2022 06:24:30.899 * Running
mode=standalone, port=6379.
embed-cache-1    | 1:M 10 Sep 2022 06:24:30.899 # Server
initialized
...
embed-cache-1    | 1:M 10 Sep 2022 06:24:30.899 * Loading RDB
```

```
produced by version 6.2.7
embed-cache-1    | 1:M 10 Sep 2022 06:24:30.899 * RDB age 10
seconds
...
embed-cache-1    | 1:M 10 Sep 2022 06:24:30.899 * Ready to
accept connections
```

Once everything is running, you should be able to access the server by opening your browser and typing `http://localhost:3333` in the address bar.

The Docker Compose file looks as follows:

```
version: '3'
services:
  server:
    build: .
    ports:
      - "3333:3333"
  cache:z
    image: redis:7.0.4-alpine
    restart: always
    ports:
      - '6379:6379'
```

The file outlines two containers that need to be run – the server is pointing to our application server, and the `build` parameter uses the `.` dot notation. This tells Docker Compose that the source (Dockerfile) to build the image for this container is found in the local directory, while the cache service is a Redis server, and it will be pulled from the Docker remote registry, specifically version 7.0.4.

Summary

In this chapter, we learned about what Docker is and how to use it. Building applications is one part of the puzzle, but packaging them to be deployed in a cloud environment requires developers to understand Docker and how to build Docker images for their applications. We looked at how Docker stores images on your local machine and also inspected the state of the running container.

We learned that when containers are running, there is a lot of information generated that can help us to understand what's going on with the container and also the parameters used to run our application. We also learned about the `Dockerfile` and used it to package our sample application into a container to run it as a single Docker image.

In the next chapter, we will use the knowledge we gained in this chapter by deploying our images to a cloud environment.

14

Cloud Deployment

In this chapter, we will learn about cloud deployment, specifically using AWS as the cloud provider. We will look at some of the infrastructure services provided by AWS and how to use them. We will learn about using and writing code for creating the different AWS infrastructure services using an open source tool called Terraform. Understanding the cloud and how cloud deployment works has become a necessity for developers nowadays rather than an exception. Gaining a good understanding of the different aspects of cloud deployment will allow you to think outside the box about how your application should run in the cloud.

Upon completion of this chapter, we will have learned about the following key topics:

- Learning basic AWS infrastructure
- Understanding and using Terraform
- Writing Terraform for local and cloud deployment
- Deploying to AWS Elastic Container Service

The end goal of this chapter is to provide you with some knowledge about the cloud and how to perform certain basic operations for deploying applications to the cloud.

Technical requirements

All the source code explained in this chapter can be checked out at `https://github.com/PacktPublishing/Full-Stack-Web-Development-with-Go/tree/main/chapter14`.

This chapter uses AWS services, so you are expected to have an AWS account. AWS provides a Free Tier for new user registration; more information can be found at `https://aws.amazon.com/free`.

> **Note**
> Using any kind of AWS services will incur a cost. Please read and inform yourself before using the service. We highly recommend reading what is available on the Free Tier on the AWS website.

AWS refresher

AWS stands for **Amazon Web Services** and belongs to Amazon, which provides the e-commerce platform amazon.com.au. AWS provides services that allow organizations to run their applications in a complete infrastructure without owning any of the hardware required.

The AWS brand is a household name for developers and almost all developers have some basic direct/indirect exposure to using AWS tools or its services. In this section, we will look at some services provided by AWS as a refresher.

The question that comes to our mind is, why bother using services such as AWS? *Figure 14.1* summarizes the answer nicely. AWS provides services that are available across different continents of the world and ready to be used by organizations to fulfill their needs. Imagine that your organization has customers across different continents. How much easier would it be to run your application on different continents without having the burden of investing in hardware on each of those continents?

Figure 14.1: Global AWS Regions

In the next section, we will look at the basic service provided by AWS called AWS EC2, which provides computing resources.

Amazon Elastic Compute Cloud

Amazon **Elastic Compute Cloud (EC2)** is the basic computing resource for developers to run their applications on. You can think of EC2 as a virtual computer on Amazon infrastructure somewhere on the internet that runs your application. You can select from a number of computer configurations that you want to run your application on, from a small 512-MB memory to a gigantic 384-GB memory computer with different configurations of storage. *Figure 14.2* shows the Instance Type Explorer that can be accessed using the following URL: `https://aws.amazon.com/ec2/instance-explorer/`.

Figure 14.2: Instance Type Explorer

In the next section, we will look at another AWS resource related to computing that is super important for applications, and that is storage.

Storage

Computing power is great for running applications, but applications require long-term storage to store data such as log files and databases. There are a number of different kinds of storage provided by AWS. For example, *Figure 14.3* shows the **Elastic Block Store** (**EBS**), which is a block storage service. This block storage is like the normal storage that you have on your local computer and is offered as a hard drive or a **solid-state drive** (**SSD**).

Figure 14.3: EBS

The amazing thing about having this kind of storage is its elastic nature – what this means is you can increase or decrease the size of storage anytime you need without the worry of adding new hardware. Imagine what would happen if you were running out of hard drive space on your local computer. You would need to buy a new hard drive and install and configure it, none of which is required when you use the AWS storage service. Attaching storage to the EC2 instance of your choice enables your application to run and store data in the cloud.

We will look at another AWS service that is as important as the one that we have just discussed: networking.

Virtual Private Cloud

Now that your application is running in its own virtual computer, complete with storage, the next question is how we configure a network in AWS so that users can access the application. This is called a **Virtual Private Cloud** (**VPC**). Think about a VPC as your own network setup, but without cables – everything is configured and run using software. *Figure 14.4* shows the powerful capability of a VPC, enabling you to connect different networks configured in different Regions.

Think of a Region as the physical location where AWS stores its hardware, and if you run your applications in different physical locations, you are able to connect them using a VPC.

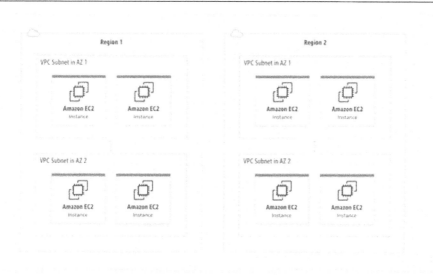

Figure 14.4: Virtual Private Networking

You have full control to configure the network of each Region your application is running on, how these Regions communicate with your own network, and how your application will be accessible via the public internet.

In the next section, we will look at another important service that a lot of applications require which is storing data in a database.

Database storage

No matter what kind of applications you are building, you will require a database to store data, and this requires a database server to be running. AWS provides different database services ranging from those that store small amounts of data to massively distributed databases across different continents. One of these services is called Amazon **Relational Database Service** (**RDS**), a managed service to set up, scale, and operate databases.

The databases that RDS can support are MySQL, PostgreSQL, MariaDB, Oracle, and SQL Server. *Figure 14.5* outlines the features provided by RDS.

Figure 14.5: RDS

Elastic Container Service

In *Chapter 13, Dockerizing an Application*, we learned how to create Docker images to package our application so it can run as a container. Packaging applications as Docker images allows us to run our application in any kind of environment, from a local machine to the cloud. AWS provides a related service called **Elastic Container Service** (**ECS**).

ECS helps us to deploy, manage, and scale out applications that have been built as containers. A key scaling feature of ECS is the ability to scale your application using the Application Auto Scaling capability. This feature allows developers to scale applications based on certain conditions, such as the following:

- **Step scaling**: This means scaling an application based on the breach of an alarm
- **Scheduled scaling**: This is scaling based on a predetermined time

AWS tools

AWS provides different ways to use its services, including a web user interface and the **command-line interface** (**CLI**). The main page of the web UI can be seen in *Figure 14.6*. You will need to register for an AWS account first before using any of the AWS tools.

The UI is a very good place to start exploring the different services and go through some sample tutorials to get a better understanding of each service.

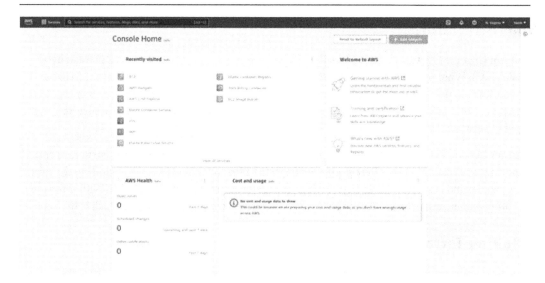

Figure 14.6: AWS web UI

The other AWS tool that is used to interact with the services is the CLI, which needs to be installed locally (`https://docs.aws.amazon.com/cli/latest/userguide/getting-started-install.html`). The CLI makes it easier to interact with the AWS services than the web UI. If you have installed it locally, when you run aws from your terminal, you will see the following output:

```
usage: aws [options] <command> <subcommand> [<subcommand> ...]
[parameters]
To see help text, you can run:


  aws help
  aws <command> help
  aws <command> <subcommand> help


 aws: error: the following arguments are required: command
```

In the next section, we will look at how to use some of the features described here to deploy our application in AWS.

Understanding and using Terraform

In this section, we will look at another tool that makes it easier for us to work with AWS services: Terraform. In the previous section, we learned that AWS provides tools of its own, which is great for small tasks, but once you start combining the different services it becomes harder to use them.

What is Terraform?

Terraform (`https://www.terraform.io/`) is an open source tool that provides **infrastructure as code** (**IaC**). What this means is you write code to define what kind of service you want to use and how you want to use it, and this way, you can combine and link the different services together as a single piece. This makes it easy for you as a developer to run and destroy infrastructure as a unit instead of separate fragments.

The other benefit that Terraform provides is the ability to version control the infrastructure code like normal application code, where it goes through the normal review process, including the peer review process and also unit testing, before deploying the infrastructure to production. With this, your application and infrastructure will now go through the same development process, which is trackable.

Installing Terraform

The Terraform installation process is straightforward: you can find a complete set of instructions for your operating system in the HashiCorp documentation at `https://www.terraform.io/downloads`.

For example, when writing this book we are using an Ubuntu-based distro, so we download the AMD64 binary from `https://releases.hashicorp.com/terraform/1.3.0/terraform_1.3.0_linux_amd64.zip` and include the Terraform directory into our `PATH`, as in the following snippet. The directory added to the `PATH` variable environment is a temporary solution for the terminal that you are using. In order to store it, you need to put it as part of your shell script (for Linux, if you are using Bash, you can add this to your `.bashrc` file):

```
export PATH=$PATH:/home/user/Downloads/
```

To test whether the installation was successful, open the terminal and execute `Terraform`:

```
Terraform
```

You should get the following output:

```
Usage: terraform [global options] <subcommand> [args]

The available commands for execution are listed below.
The primary workflow commands are given first, followed by
less common or more advanced commands.

Main commands:
  init            Prepare your working directory for other
                  commands
```

```
    ...

All other commands:
    console        Try Terraform expressions at an interactive
                   command prompt
    fmt            Reformat your configuration in the standard
                   style
    ...
```

For detailed information on how to install Terraform for your environment, see `https://developer.hashicorp.com/terraform/tutorials/aws-get-started/install-cli`.

Now that we have completed the Terraform installation, we will learn how to use some of the basic commands available in Terraform. The commands will enable you to jumpstart your journey into the world of cloud deployment.

Terraform basic commands

In this section, we will learn some basic Terraform commands that are often used when writing code. We will also examine concepts that are relevant to Terraform.

The init command

Every time we start writing Terraform code, the first command that we run is `terraform init`. This command prepares all the necessary dependencies required to run the code locally. The command performs the following steps:

1. Downloads all the necessary modules that are used in the code.

2. Initializes plugins that are used in the code. For example, if the code is deployed on AWS it will download the AWS plugins.

3. Creates a file called a lock file that registers the different dependencies and versions that are used by the code.

To gain a better understanding of the previous steps, let's run the command. Open the terminal and change to the `chapter14/simple` directory, and execute the following command:

```
terraform init
```

You will see an output as follows:

```
Initializing the backend...
Initializing provider plugins...
```

```
- Finding kreuzwerker/docker versions matching "~> 2.16.0"...
- Installing kreuzwerker/docker v2.16.0...
- Installed kreuzwerker/docker v2.16.0 (self-signed, key ID
BD080C4571C6104C)
```

```
...
```

Once the `init` process is complete, your directory will look like the following:

```
.
├── main.tf
├── .terraform
│   └── providers
│       └── registry.terraform.io
│           └── kreuzwerker
│               └── docker
│                   └── 2.16.0
│                       └── linux_amd64
│                           ├── CHANGELOG.md
│                           ├── LICENSE
│                           ├── README.md
│                           └── terraform-provider-docker_
v2.16.0
├── .terraform.lock.hcl
└── versions.tf
```

The `.terraform` directory contains the dependencies that are specified in the code. In this example, it uses the `kreuzwerker/docker` plugin, which is used to run Docker containers.

The `.terraform.lock.hcl` file contains the version information of the dependencies, and it looks like the following:

```
# This file is maintained automatically by "terraform
# init".
# Manual edits may be lost in future updates.

provider "registry.terraform.io/kreuzwerker/docker" {
  version     = "2.16.0"
  constraints = "~> 2.16.0"
  hashes = [
```

```
    "h1:OcTn2QyCQNjDiJYy1vqQFmz2dxJdOF/2/HBXBvGxU2E=",
    ...
  ]
}
```

The plan command

The plan command is used to help us understand the execution plan that Terraform will be doing. This is a very important feature as it gives us visibility of what changes will be performed to our infrastructure. This will give us a better understanding of which parts of the infrastructure will be impacted by the code. Unlike tools such as Chef or Ansible, Terraform is interesting in that it will tend towards a target state and only make the changes necessary to reach it. For example, if you had a target of five EC2 instances but Terraform only knew of three, it would take the steps needed to reach that target of five.

Open the terminal, change to the chapter14/simple directory, and execute the following command:

```
terraform plan
```

You will get the following output:

```
...
Terraform will perform the following actions:

  # docker_container.nginx will be created
  + resource "docker_container" "nginx" {
      + attach           = false
      + bridge           = (known after apply)
      + command          = (known after apply)
      + container_logs   = (known after apply)
      + entrypoint       = (known after apply)
      + env              = (known after apply)
      + exit_code        = (known after apply)
      ...
      + remove_volumes   = true
      + restart          = "no"
      + rm               = false
      + security_opts    = (known after apply)
      + shm_size         = (known after apply)
      + start            = true
```

```
        + stdin_open        = false
        + tty               = false

        + healthcheck {
            + interval      = (known after apply)
            + retries       = (known after apply)
            + start_period  = (known after apply)
            + test          = (known after apply)
            + timeout       = (known after apply)
        }

        + labels {
            + label = (known after apply)
            + value = (known after apply)
        }

        + ports {
            + external = 8000
            + internal = 80
            + ip       = "0.0.0.0"
            + protocol = "tcp"
        }
    }

  # docker_image.nginx will be created
  + resource "docker_image" "nginx" {
        + id            = (known after apply)
        ...
        + repo_digest   = (known after apply)
    }

Plan: 2 to add, 0 to change, 0 to destroy.
...
```

The output shows that there will be 2 things added and 0 operations for changing or destroying, which tells us that this is the first time we are running the code or it's still fresh.

The apply command

The normal process of running Terraform is that after `init`, we run `apply` (however, if we are not sure about the impact, we use the `plan` command as shown previously). Open the terminal, change to the `chapter14/simple` directory, and execute the following command:

```
terraform apply –auto-aprove
```

You will get the following output:

```
...
Terraform will perform the following actions:

  # docker_container.nginx will be created
  + resource "docker_container" "nginx" {
      + attach          = false
      + bridge          = (known after apply)
      ...
    }


  # docker_image.nginx will be created
  + resource "docker_image" "nginx" {
      + id              = (known after apply)
      ...
    }


Plan: 2 to add, 0 to change, 0 to destroy.
docker_image.nginx: Creating...
docker_image.nginx: Still creating... [10s elapsed]
docker_image.nginx: Creation complete after 17s
[id=sha256:2d389e545974d4a93ebdef09b650753a55f72d1ab4518d17a
30c0e1b3e297444nginx:latest]
docker_container.nginx: Creating...
docker_container.nginx: Creation complete after 2s [id=d0c94bd4
b548e6a19c3afb907a777bcb602e965bc05db8ef6d0d380601bb0694]


Apply complete! Resources: 2 added, 0 changed, 0 destroyed.
```

As seen in the output, the `nginx` container will be downloaded (if it does not exist as yet) and then run. Once the command is successfully run you can test it by opening your browser and accessing `http://localhost:8080`. You will see something like *Figure 14.7*.

Welcome to nginx!

If you see this page, the nginx web server is successfully installed and working. Further configuration is required.

For online documentation and support please refer to nginx.org.
Commercial support is available at nginx.com.

Thank you for using nginx.

Figure 14.7: nginx running in a container

The destroy command

The last command that we will look at is `destroy`. As the name implies, it is used to destroy the infrastructure that was created using the `apply` command. Use this command with caution if you are unsure about the impact of the code on your infrastructure. Use the `plan` command before running this to get better visibility of what will be removed from the infrastructure.

Open the terminal and run the following command from the `chapter14/simple` directory:

```
Terraform destroy -auto-approve
```

You will get the following output:

```
docker_image.nginx: Refreshing state... [id=sha256:
2d389e545974d4a93ebdef09b650753a55f72d1ab4518d17a30c
0e1b3e297444nginx:latest]
docker_container.nginx: Refreshing state... [id=9c46cff8
1a27edb6aba08a448d715599c644aaa79b192728016db0d903da9fb0]


...


Terraform will perform the following actions:

  # docker_container.nginx will be destroyed
  - resource "docker_container" "nginx" {
      - attach            = false -> null
      - command           = [
          - "nginx",
```

```
          - "-g",
          - "daemon off;",
        ] -> null
      - cpu_shares          = 0 -> null

        ...
    }


  # docker_image.nginx will be destroyed
  - resource "docker_image" "nginx" {
      - id           =
          "sha256:2d389e545974d4a93ebdef09b650753a55f7
          2d1ab4518d17a30c0e1b3e297444nginx:latest" ->
          null
      - keep_locally = false -> null
      - latest       =
          "sha256:2d389e545974d4a93ebdef09b650753a55f72
          d1ab4518d17a30c0e1b3e297444" -> null
      - name         = "nginx:latest" -> null
      - repo_digest  =
          "nginx@sha256:0b970013351304af46f322da126351
          6b188318682b2ab1091862497591189ff1" -> null
    }


Plan: 0 to add, 0 to change, 2 to destroy.
docker_container.nginx: Destroying... [id=9c46cff81a27edb6aba
08a448d715599c644aaa79b192728016db0d903da9fb0]
docker_container.nginx: Destruction complete after 1s
docker_image.nginx: Destroying... [id=sha256:2d389e545974d4a93
ebdef09b650753a55f72d1ab4518d17a30c0e1b3e297444nginx:latest]
docker_image.nginx: Destruction complete after 0s


Destroy complete! Resources: 2 destroyed.
```

In the output, we can see that there are 2 infrastructures that are destroyed – one is the container removed from memory, and the other is the removal of the image from the local Docker registry.

The -auto-approve command is used to automatically approve the steps; normally, without using this, Terraform will stop execution and ask the user to enter Yes or No to continue at each step. This is a precautionary measure to ensure that the user does indeed want to destroy the infrastructure.

In the next section, we will look at writing Terraform code and how it uses providers. We will look at a few Terraform examples to get an understanding of how it works to spin up different AWS infrastructure services for deploying applications.

Coding in Terraform

HashiCorp, the creator of Terraform, created **HashiCorp configuration language** (**HCL**), which is used in writing Terraform code. HCL is a functional programming language with features such as loops, if statements, variables, and logic flow that are normally found in programming languages. Complete in-depth HCL documentation can be found at https://www.terraform.io/language/.

Providers

The reason why Terraform is so widely used is the number of extensions that are available from the company and open source communities; these extensions are called providers. A provider is a piece of software that interacts with the different cloud providers and other resources in the cloud. We will look at Terraform code to understand more about providers. The following code snippets can be found inside the chapter14/simple directory:

```
terraform {
 required_providers {
   docker = {
     source = "kreuzwerker/docker"
     version = "~> 2.16.0"
   }
 }
}

resource "docker_image" "nginx" {
 name          = "nginx:latest"
 keep_locally = false
}

resource "docker_container" "nginx" {
 image = docker_image.nginx.name
 name  = "hello-terraform"
```

```
ports {
  internal = 80
  external = 8000
}
}
```

The `resource` block in the code can be used to declare infrastructure or an API. In this example, we are using Docker, specifically, `docker_image` and `docker_container`. When Terraform runs the code it detects the `required_providers` block, which is used to define a provider. A provider is an external module that the code will be using, and this will be automatically downloaded by Terraform from a central repository. In our example, the provider that we are using is the `kreuzwerker/docker` Docker provider. More information on this provider can be found at the following link: `https://registry.terraform.io/providers/kreuzwerker/docker/`.

Open the terminal, make sure you are inside the `chapter14/simple` directory, and run the following command:

```
terraform init
```

You will see the following output in your terminal:

```
Initializing the backend...

Initializing provider plugins...
- Finding kreuzwerker/docker versions matching "~> 2.16.0"...
- Installing kreuzwerker/docker v2.16.0...
- Installed kreuzwerker/docker v2.16.0 (self-signed, key ID
BD080C4571C6104C)
...
```

Terraform downloads the provider and stores it inside the `chapter14/simple/.terraform` folder. Now, let's run the sample code and see what we get, by running the following command in the same terminal:

```
terraform apply -auto-approve
```

You will see the following output:

```
...
  # docker_container.nginx will be created
  + resource "docker_container" "nginx" {
      + attach          = false
```

```
        ...
    }

    # docker_image.nginx will be created
    + resource "docker_image" "nginx" {
        + id              = (known after apply)
        ...
    }

Plan: 2 to add, 0 to change, 0 to destroy.
    ...
docker_image.nginx: Creation complete after 22s
[id=sha256:2d389e545974d4a93ebdef09b650753a55f72d1ab4518d17a
30c0e1b3e297444nginx:latest]
docker_container.nginx: Creating...

docker_container.nginx: Creation complete after 2s [id=b860780
af83a4c719a916b87171d96801cc2243a61242354815f6d82dc6a5e40]
```

Open your browser and go to `http://localhost:8000`. You will see something like *Figure 14.7*.

Terraform downloads the `nginx` Docker image automatically to your local machine and runs the `nginx` container using the port defined in the `ports` code block (port `8000`). To destroy the running container and delete the image locally from the Docker registry, all you have to do is run the following command:

```
terraform destroy -auto-approve
```

If you compare the steps involved to do the same thing manually using the Docker command, it is more involved and error-prone; writing it in Terraform makes it much easier to run and remove containers with a single command.

In the next section, we will explore more examples to better understand how to use Terraform for deploying applications.

Terraform examples

In the following sections, we will look at different ways we can use Terraform, such as pulling images from GitHub and running them locally, or building and publishing Docker images.

> **Note**
>
> Make sure every time you run Terraform examples that create AWS resources to remember to destroy the resources using the `terraform destroy` command.
>
> All resources created in AWS incur charges, and by destroying them, you will ensure there will be no surprise charges.

Pulling from GitHub Packages

The example code for this section can be found inside the `chapter14/github` folder. The following snippet is from `pullfromgithub.tf`:

```
#script to pull chapter12 image and run it locally
#it also store the image locally
terraform {
  required_providers {
    docker = {
      source  = "kreuzwerker/docker"
      version = "~> 2.13.0"
    }
  }
}



data "docker_registry_image" "github" {
 name = "ghcr.io/nanikjava/golangci/chapter12:latest"
}


resource "docker_image" "embed" {
  ...
}


resource "docker_container" "embed" {
  ...
}
```

The main objective of the code is to download the Docker image that we built in *Chapter 12, Building Continuous Integration*. Once the Docker image is downloaded, it will be run locally. Open your terminal, make sure you are inside the `chapter14/github` directory, and run the following command:

```
terraform init
```

Then run the following command:

```
terraform apply -auto-approve
```

You will see output in your terminal that looks like the following:

```
...
data.docker_registry_image.github: Reading...
data.docker_registry_image.github: Read complete after 1s
[id=sha256:a355f55c33a400290776faf20b33d45096eb19a6431fb
0b3f723c17236e8b03e]
...


  # docker_container.embed will be created
  + resource "docker_container" "embed" {
      + attach          = false
    ...


      + ports {
          + external = 3333
          + internal = 3333
          ...
      }
  }


  # docker_image.embed will be created
  + resource "docker_image" "embed" {
      ...
      + name           =
        "ghcr.io/nanikjava/golangci/chapter12:latest"
      ...
  }


Plan: 2 to add, 0 to change, 0 to destroy.
... [id=sha256:684e34e77f40ee1e75bfd7d86982a4f4fae1dbea3286682af
3222a270faa49b7ghcr.io/nanikjava/golangci/chapter12:latest]
docker_container.embed: Creation complete after 7s
```

```
[id=f47d1ab90331dd8d6dd677322f00d9a06676b71dda3edf2cb2e66
edc97748329]


Apply complete! Resources: 2 added, 0 changed, 0 destroyed.
```

Open your browser and go to `http://localhost:3333`. You will see the login page of the sample app.

The code uses the same `docker` provider that we discussed in the previous section, and we use a new `docker_registry_image` command to specify the address to download the Docker image from, in this case from the `ghcr.io/nanikjava/golangci/chapter12:latest` GitHub package.

The other HCL feature we are using is the `data` block, as shown here:

```
...
data "docker_registry_image" "github" {
  name = "ghcr.io/nanikjava/golangci/chapter12:latest"
}
...
```

The `data` block works similarly to `resource`, except it is only used for reading values and not creating or destroying resources or to get data that will be used internally as configuration to another resource. In our sample, it is used by the `docker_image` resource, as shown here:

```
resource "docker_image" "embed" {
  keep_locally = true
  name         = "${data.docker_registry_image.github.name}"
}
```

AWS EC2 setup

In the previous examples, we looked at using the Docker provider to run Docker containers locally. In this example, we will look at creating AWS resources, specifically EC2 instances. An EC2 instance is basically a virtual machine that can be initialized with a certain configuration to run in the cloud to host your application.

In order to create resources in AWS, you will first need to already have an AWS account. If you don't have an AWS account, you can create one at `https://aws.amazon.com/`. Once you have your AWS account ready, log in to the AWS website, and in the main console (*Figure 14.6*) web page, click on your name on the right side and it will display a drop-down menu, as shown in *Figure 14.8*. Then click on **Security credentials**.

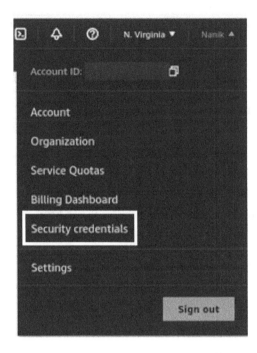

Figure 14.8: Security credentials option

Your browser will now show the **identity and access management** (**IAM**) page, as shown in *Figure 14.9*. Select the **Access keys (access key ID and secret access key)** option. Since you haven't created any key, it will be empty. Click on the **Create New Access Key** button and follow the instructions to create a new key.

Figure 14.9: Access keys section

Once you complete the steps you will get two keys – an Access Key ID and Secret Access Key. Keep these keys safe as they are used like a username and password combination you use to create resources in AWS infrastructure.

Now that you have the keys required, you can now open a terminal and change into the `chapter14/simpleec2` directory, and run the example as follows:

```
terraform init
```

Next, run the following command to create the EC2 instance:

```
terraform apply  -var="aws_access_key=xxxx" -var="aws_secret_key=xxx" -auto-approve
```

Once completed you will see the output as follows:

```
...
Terraform will perform the following actions:

  # aws_instance.app_server will be created
  + resource "aws_instance" "app_server" {
     + ami = "ami-0ff8a91507f77f867"
     ...
    }


  # aws_subnet.default-subnet will be created
  + resource "aws_subnet" "default-subnet" {
     ...
    }


  # aws_vpc.default-vpc will be created
  + resource "aws_vpc" "default-vpc" {
     + arn                       = (known after apply)
     ...
    }

Plan: 3 to add, 0 to change, 0 to destroy.
...
aws_instance.app_server: Creation complete after 24s [id=i-0358d1df58e055d70]
```

The output shows three resources were created – the AWS instance (EC2), an IP subnet, and a network VPC. Now, let's take a look at the code (the complete code can be seen inside the `chapter14/simpleec2` directory). The code requires your AWS keys, storing them inside the `variable` block as `aws_access_key` and `aws_secret_key`:

```
terraform {
  ...
}


variable "aws_access_key" {
  type = string
}
variable "aws_secret_key" {
  type = string
}


provider "aws" {
  region     = "us-east-1"
  access_key = var.aws_access_key
  secret_key = var.aws_secret_key
}
```

The keys will be passed to the `aws` provider to enable the provider to communicate with the AWS service using our keys.

The following part of the code creates the VPC and IP subnet, which will be used as a private network by EC2 instances:

```
resource "aws_vpc" "default-vpc" {
  cidr_block           = "10.0.0.0/16"
  enable_dns_hostnames = true
  tags                 = {
    env = "dev"
  }
}
resource "aws_subnet" "default-subnet" {
  cidr_block = "10.0.0.0/24"
  vpc_id     = aws_vpc.default-vpc.id
}
```

The last resource the code defines is the EC2 instance, as follows:

```
resource "aws_instance" "app_server" {
  ami             = "ami-0ff8a91507f77f867"
  instance_type   = "t2.nano"
  subnet_id       = aws_subnet.default-subnet.id

  tags = {
    Name = "Chapter14"
  }
}
```

The EC2 instance type is t2.nano, which is the smallest virtual machine that can be configured. It is linked to the IP subnet defined earlier by assigning the subnet ID to the subnet_id parameter.

Deploying to ECS with a load balancer

The last example that we are going to look at is using AWS ECS. The source code can be found inside the chapter14/lbecs directory. The code will use ECS to deploy our *Chapter 12* container hosted in GitHub Packages and made scalable by using a load balancer. *Figure 14.9* shows the infrastructure configuration after running the code.

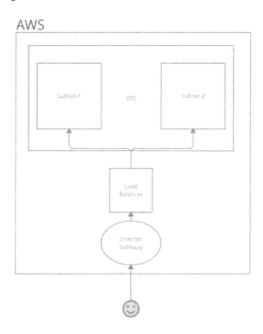

Figure 14.10: ECS with a load balancer

The code uses the following services:

- **An internet gateway**: As the name implies, this is a gateway that enables communication to be established between the AWS VPC private network and the internet. With the help of the gateway, we open our application to the world.

- **A load balancer**: This service helps balance the incoming traffic across the different networks configured, ensuring that the application can take care of all incoming requests.

ECS provides the capability to scale the deployment process for containers. This means that, as developers, we don't have to worry about how to scale the containers that are running our application, as this is all taken care of by ECS. More in-depth information can be found at `https://aws.amazon.com/ecs/`. The application is run the same way as in the previous examples, using the `terraform init` and `terraform apply` commands.

> **Note**
> The ECS example takes a bit longer to execute compared to the other examples.

You will get output that looks like the following:

```
...
Terraform will perform the following actions:

  # aws_default_route_table.lbecs-subnet-default-route-
  # table will be created
  + resource "aws_default_route_table"
            "lbecs-subnet-default-route-table" {
      ...
    }


  # aws_ecs_cluster.lbecs-ecs-cluster will be created
  + resource "aws_ecs_cluster" "lbecs-ecs-cluster" {
      ...
    }


  # aws_ecs_service.lbecs-ecs-service will be created
  + resource "aws_ecs_service" "lbecs-ecs-service" {
      ...
    }
```

```
# aws_ecs_task_definition.lbecs-ecs-task-definition will
# be created
+ resource "aws_ecs_task_definition"
          "lbecs-ecs-task-definition" {
    ...
  }


# aws_internet_gateway.lbecs-igw will be created
+ resource "aws_internet_gateway" "lbecs-igw" {
    ...
  }


# aws_lb.lbecs-load-balancer will be created
+ resource "aws_lb" "lbecs-load-balancer" {
    ...
  }


# aws_lb_listener.lbecs-load-balancer-listener will be
# created
+ resource "aws_lb_listener"
          "lbecs-load-balancer-listener" {
    ...
  }


# aws_lb_target_group.lbecs-load-balancer-target-group
# will be created
+ resource "aws_lb_target_group"
          "lbecs-load-balancer-target-group" {
    ...
  }


# aws_security_group.lbecs-security-group will be created
+ resource "aws_security_group" "lbecs-security-group" {
    ...
  }
```

```
    # aws_subnet.lbecs-subnet will be created
    + resource "aws_subnet" "lbecs-subnet" {
        ...
      }

    # aws_subnet.lbecs-subnet-1 will be created
    + resource "aws_subnet" "lbecs-subnet-1" {
        ...
      }

    # aws_vpc.lbecs-vpc will be created
    + resource "aws_vpc" "lbecs-vpc" {
        ...
      }

Plan: 12 to add, 0 to change, 0 to destroy.

...
aws_ecs_service.lbecs-ecs-service: Creation complete after
2m49s [id=arn:aws:ecs:us-east-1:860976549008:service/lbecs-ecs-
cluster/lbecs-ecs-service]

...

Outputs:

url = "load-balancer-1956367690.us-east-1.elb.amazonaws.com"
```

Let's break down the code to see how it uses ECS and configures the internet gateway, load balancer, and network. The following code shows the internet gateway declaration, which is simple enough as it requires to be attached to a VPC:

```
resource "aws_internet_gateway" "lbecs-igw" {
  vpc_id = aws_vpc.lbecs-vpc.id

  tags = {
```

```
    Name = "Internet Gateway"
  }
}


resource "aws_default_route_table" "lbecs-subnet-default-route-
table" {
  default_route_table_id =
    aws_vpc.lbecs-vpc.default_route_table_id
  route {
    cidr_block = "0.0.0.0/0"
    gateway_id = "${aws_internet_gateway.lbecs-igw.id}"
  }
}
```

Besides that, the gateway will also be attached to a routing table declared inside the `aws_default_route_table` block. This is necessary as this tells the gateway how to route the incoming and outgoing traffic through the internal private VPC network.

Now that our internal private network can communicate to the internet via a gateway, we need to have network rules in place to ensure our network is secure, and this is done in the following code:

```
resource "aws_security_group" "lbecs-security-group" {
  name        = "allow_http"
  description = "Allow HTTP inbound traffic"
  vpc_id      = aws_vpc.lbecs-vpc.id

  egress {
    from_port   = 0
    to_port     = 0
    protocol    = "-1"
    cidr_blocks = ["0.0.0.0/0"]
  }

  ingress {
    description = "Allow HTTP for all"
    from_port   = 80
    to_port     = 3333
    protocol    = "tcp"
```

```
    cidr_blocks = ["0.0.0.0/0"]
  }
}
```

The `egress` block declares the rule for outgoing network traffic, allowing all protocols to pass through. The incoming network traffic rule is declared in the `ingress` block, and allows ports between `80-3333` and only over TCP.

Using a load balancer requires two different subnets to be declared. In our code example, this is as follows:

```
resource "aws_lb" "lbecs-load-balancer" {
  name               = "load-balancer"
  internal           = false
  load_balancer_type = "application"
  security_groups    = [aws_security_group.lbecs-security-group.
                         id]
  subnets            = [aws_subnet.lbecs-subnet.id,
                          aws_subnet.lbecs-subnet-1.id]
  tags               = {
    env = "dev"
  }
}
```

The last piece of code that we will look at is the ECS block, as follows:

```
resource "aws_ecs_cluster" "lbecs-ecs-cluster" {
  name = "lbecs-ecs-cluster"
}


resource "aws_ecs_task_definition" "lbecs-ecs-task-definition"
{
  family                   = "service"
  requires_compatibilities = ["FARGATE"]
  network_mode             = "awsvpc"
  cpu                      = 1024
  memory                   = 2048
  container_definitions    = jsonencode([
    {
      name        = "lbecs-ecs-cluster-chapter14"
      image       =
```

```
        "ghcr.io/nanikjava/golangci/chapter12:latest"
      ...

      portMappings = [
        {
          containerPort = 3333
        }
      ]
    }
 ])
}


resource "aws_ecs_service" "lbecs-ecs-service" {
  name            = "lbecs-ecs-service"
  cluster         = aws_ecs_cluster.lbecs-ecs-cluster.id
  task_definition =
    aws_ecs_task_definition.lbecs-ecs-task-definition.arn
  desired_count   = 1
  launch_type     = "FARGATE"

  network_configuration {
    ...

  }


  load_balancer {
    target_group_arn = aws_lb_target_group.lbecs-load-
                       balancer-target-group.arn
    container_name   = "lbecs-ecs-cluster-chapter14"
    container_port   = 3333
  }

  tags = {
    env = "dev"
  }
}
```

The preceding code contains three different code blocks that are explained as follows:

- `aws_ecs_cluster`: This block configures the name of the ECS cluster
- `aws_ecs_task_definition`: This block configures the ECS task, which specifies what kind of container it has to run, the virtual machine configuration that the container will be running on, the network mode, security group, and other options
- `aws_ecs_service`: This block ties together the different services to describe the complete infrastructure that will be run, such as security, ECS task, network configuration, load balancers, public IP address, and more

Once ECS has been spun up, it will print out in your console the load-balanced public address you can use to access the application. For example, when it was run, we got the following output in the terminal:

```
...
aws_lb_listener.lbecs-load-balancer-listener: Creating...
aws_lb_listener.lbecs-load-balancer-listener: Creation
complete after 1s [id=arn:aws:elasticloadbalancing:us-east-
1:860976549008:listener/app/load-balancer/4ad0f8b815a06f02/
d945bba078d0c365]
aws_ecs_service.lbecs-ecs-service: Creation complete after
2m27s [id=arn:aws:ecs:us-east-1:860976549008:service/lbecs-ecs-
cluster/lbecs-ecs-service]

Apply complete! Resources: 12 added, 0 changed, 0 destroyed.

Outputs:

url = "load-balancer-375816308.us-east-1.elb.amazonaws.com"
```

Using the `load-balancer-375816308.us-east-1.elb.amazonaws.com` address in the browser will show the application login page. This address is dynamically generated by AWS, and you will get something different than what is shown in the previous output.

Summary

In this chapter, we explored cloud solutions provided by AWS, and we briefly looked at the different services offered, such as EC2, VPC, storage, and others. We learned about the open source Terraform tools that make it easy to create, manage, and destroy cloud infrastructure in AWS.

We learned how to install and use Terraform locally, and how to write Terraform code to use Docker as a provider, allowing us to run containers locally. Terraform also allows us to download, run, and destroy containers locally with a single command.

We also explored different Terraform examples for creating AWS infrastructure resources and looked at one of the advanced features of AWS ECS.

In this last chapter of the book, you have learned the different things that need to be done to deploy an application to the AWS cloud.

Index

www.packtpub.com

Subscribe to our online digital library for full access to over 7,000 books and videos, as well as industry leading tools to help you plan your personal development and advance your career. For more information, please visit our website.

Why subscribe?

- Spend less time learning and more time coding with practical eBooks and Videos from over 4,000 industry professionals

- Improve your learning with Skill Plans built especially for you

- Get a free eBook or video every month

- Fully searchable for easy access to vital information

- Copy and paste, print, and bookmark content

Did you know that Packt offers eBook versions of every book published, with PDF and ePub files available? You can upgrade to the eBook version at packtpub.com and as a print book customer, you are entitled to a discount on the eBook copy. Get in touch with us at customercare@packtpub.com for more details.

At www.packtpub.com, you can also read a collection of free technical articles, sign up for a range of free newsletters, and receive exclusive discounts and offers on Packt books and eBooks.

Other Books You May Enjoy

If you enjoyed this book, you may be interested in these other books by Packt:

Network Automation with Go

Nicolas Leiva, Michael Kashin

ISBN: 978-1-80056-092-5

- Understand Go programming language basics via network-related examples
- Find out what features make Go a powerful alternative for network automation
- Explore network automation goals, benefits, and common use cases
- Discover how to interact with network devices using a variety of technologies
- Integrate Go programs into an automation framework
- Take advantage of the OpenConfig ecosystem with Go
- Build distributed and scalable systems for network observability

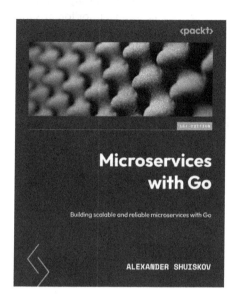

Microservices with Go

Alexander Shuiskov

ISBN: 978-1-80461-700-7

- G et familiar with the industry's best practices and solutions in microservice development
- Understand service discovery in the microservices environment
- Explore reliability and observability principles
- Discover best practices for asynchronous communication
- Focus on how to write high-quality unit and integration tests in Go applications
- Understand how to profile Go microservices

Packt is searching for authors like you

If you're interested in becoming an author for Packt, please visit `authors.packtpub.com` and apply today. We have worked with thousands of developers and tech professionals, just like you, to help them share their insight with the global tech community. You can make a general application, apply for a specific hot topic that we are recruiting an author for, or submit your own idea.

Hi!

Nick and Nanik here, authors of *Full-Stack Web Development with Go*, really hope you enjoyed reading this book and found it useful for increasing your productivity and efficiency in building and shipping production ready apps with Golang and Vue.

It would really help us (and other potential readers!) if you could leave a review on Amazon sharing your thoughts on the book.

Go to the link below or scan the QR code to leave your review:

```
https://packt.link/r/1803234199
```

Your review will help us to understand what's worked well in this book, and what could be improved upon for future editions, so it really is appreciated.

Best wishes,

NANIK

Download a free PDF copy of this book

Thanks for purchasing this book!

Do you like to read on the go but are unable to carry your print books everywhere? Is your eBook purchase not compatible with the device of your choice?

Don't worry, now with every Packt book you get a DRM-free PDF version of that book at no cost.

Read anywhere, any place, on any device. Search, copy, and paste code from your favorite technical books directly into your application.

The perks don't stop there, you can get exclusive access to discounts, newsletters, and great free content in your inbox daily

Follow these simple steps to get the benefits:

1. Scan the QR code or visit the link below

https://packt.link/free-ebook/9781803234199

2. Submit your proof of purchase
3. That's it! We'll send your free PDF and other benefits to your email directly

www.ingramcontent.com/pod-product-compliance
Lightning Source LLC
Chambersburg PA
CBHW062112050326
40690CB00016B/3290